W9-BUO-593

KAPLAN

K12 LEARNING SERVICES

Kaplan
ADVANTAGE
ACT*
Mathematics

LEVEL HS

*ACT® is a registered trademark of ACT, Inc., which does not endorse this product.

Curriculum Development
Ray Ojserkis

Contributing Writers
Gina Allison, Johanna Bunn, Karen Edwards, Adam Marks, Ingrid Multhopp,
Rachel Schwartz, Hannah Zwiebel

Design
Maurice Kessler

Production
Maria Bondoc, Gabrielle Crowley, Melissa Garcia, Scott Rayow, Stephanie Rodriguez, Lara Smiley

Illustration
Thomas Kurzanski, Erika Quiroz

Cover Design
Maurice Kessler

Production Manager
Dan Blair

Managing Editor
Michelle K. Winberg

Editorial
Robert Vinieri, Christina Castelli

Project Coordinators
Kirby Fields, Sandra Ogle

Associate Director of Mathematics Curriculum
Mark Healy

Director of Curriculum and Instruction
Deborah Lerman

Copyright © 2006 by Kaplan, Inc.

All rights reserved. No part of this book may be reproduced or transmitted in any form or by any
means, electronic or mechanical, including photocopying, recording, or by any information storage and
retrieval system, without the written permission of the Publisher, except where permitted by law.

TABLE OF CONTENTS

Unit 1
Getting to Know the ACT

© 2006 Kaplan, Inc.

Welcome to *ACT Advantage*. This program will help you prepare for the ACT Mathematics Test, which is the second of the four ACT tests you will take on Test Day. The first step in your climb to the summit of test success is the Practice Test.

This is not the actual test, so don't worry! This is simply a chance to practice skills that will be on the real test. It will also give you and your teacher a sense of your strengths and weaknesses so that you can better prepare for Test Day. Your performance will not be graded.

Like the real ACT Mathematics Test, this test has 60 problems to be completed in a total of 60 minutes. The problems cover the wide range of topics that you will see on Test Day, so that you may familiarize yourself with the subject matter of the test.

All of the questions are multiple choice. Use the answer grid to mark each answer so you can become accustomed to doing so on Test Day.

If you finish all 60 problems before time is called, review your work.

© 2006 Kaplan, Inc.

Practice Test 1

When your teacher tells you, carefully tear out this page. Then begin working.

1. Ⓐ Ⓑ Ⓒ Ⓓ Ⓔ

2. Ⓕ Ⓖ Ⓗ Ⓙ Ⓚ

3. Ⓐ Ⓑ Ⓒ Ⓓ Ⓔ

4. Ⓕ Ⓖ Ⓗ Ⓙ Ⓚ

5. Ⓐ Ⓑ Ⓒ Ⓓ Ⓔ

6. Ⓕ Ⓖ Ⓗ Ⓙ Ⓚ

7. Ⓐ Ⓑ Ⓒ Ⓓ Ⓔ

8. Ⓕ Ⓖ Ⓗ Ⓙ Ⓚ

9. Ⓐ Ⓑ Ⓒ Ⓓ Ⓔ

10. Ⓕ Ⓖ Ⓗ Ⓙ Ⓚ

11. Ⓐ Ⓑ Ⓒ Ⓓ Ⓔ

12. Ⓕ Ⓖ Ⓗ Ⓙ Ⓚ

13. Ⓐ Ⓑ Ⓒ Ⓓ Ⓔ

14. Ⓕ Ⓖ Ⓗ Ⓙ Ⓚ

15. Ⓐ Ⓑ Ⓒ Ⓓ Ⓔ

16. Ⓕ Ⓖ Ⓗ Ⓙ Ⓚ

17. Ⓐ Ⓑ Ⓒ Ⓓ Ⓔ

18. Ⓕ Ⓖ Ⓗ Ⓙ Ⓚ

19. Ⓐ Ⓑ Ⓒ Ⓓ Ⓔ

20. Ⓕ Ⓖ Ⓗ Ⓙ Ⓚ

21. Ⓐ Ⓑ Ⓒ Ⓓ Ⓔ

22. Ⓕ Ⓖ Ⓗ Ⓙ Ⓚ

23. Ⓐ Ⓑ Ⓒ Ⓓ Ⓔ

24. Ⓕ Ⓖ Ⓗ Ⓙ Ⓚ

25. Ⓐ Ⓑ Ⓒ Ⓓ Ⓔ

26. Ⓕ Ⓖ Ⓗ Ⓙ Ⓚ

27. Ⓐ Ⓑ Ⓒ Ⓓ Ⓔ

28. Ⓕ Ⓖ Ⓗ Ⓙ Ⓚ

29. Ⓐ Ⓑ Ⓒ Ⓓ Ⓔ

30. Ⓕ Ⓖ Ⓗ Ⓙ Ⓚ

31. Ⓐ Ⓑ Ⓒ Ⓓ Ⓔ

32. Ⓕ Ⓖ Ⓗ Ⓙ Ⓚ

33. Ⓐ Ⓑ Ⓒ Ⓓ Ⓔ

34. Ⓕ Ⓖ Ⓗ Ⓙ Ⓚ

35. Ⓐ Ⓑ Ⓒ Ⓓ Ⓔ

36. Ⓕ Ⓖ Ⓗ Ⓙ Ⓚ

37. Ⓐ Ⓑ Ⓒ Ⓓ Ⓔ

38. Ⓕ Ⓖ Ⓗ Ⓙ Ⓚ

39. Ⓐ Ⓑ Ⓒ Ⓓ Ⓔ

40. Ⓕ Ⓖ Ⓗ Ⓙ Ⓚ

41. Ⓐ Ⓑ Ⓒ Ⓓ Ⓔ

42. Ⓕ Ⓖ Ⓗ Ⓙ Ⓚ

© 2006 Kaplan, Inc.

43. Ⓐ Ⓑ Ⓒ Ⓓ Ⓔ   **59.** Ⓐ Ⓑ Ⓒ Ⓓ Ⓔ

44. Ⓕ Ⓖ Ⓗ Ⓙ Ⓚ   **60.** Ⓕ Ⓖ Ⓗ Ⓙ Ⓚ

45. Ⓐ Ⓑ Ⓒ Ⓓ Ⓔ

46. Ⓕ Ⓖ Ⓗ Ⓙ Ⓚ

47. Ⓐ Ⓑ Ⓒ Ⓓ Ⓔ

48. Ⓕ Ⓖ Ⓗ Ⓙ Ⓚ

49. Ⓐ Ⓑ Ⓒ Ⓓ Ⓔ

50. Ⓕ Ⓖ Ⓗ Ⓙ Ⓚ

51. Ⓐ Ⓑ Ⓒ Ⓓ Ⓔ

52. Ⓕ Ⓖ Ⓗ Ⓙ Ⓚ

53. Ⓐ Ⓑ Ⓒ Ⓓ Ⓔ

54. Ⓕ Ⓖ Ⓗ Ⓙ Ⓚ

A

55. Ⓐ Ⓑ Ⓒ Ⓓ Ⓔ

56. Ⓕ Ⓖ Ⓗ Ⓙ Ⓚ

57. Ⓐ Ⓑ Ⓒ Ⓓ Ⓔ

58. Ⓕ Ⓖ Ⓗ Ⓙ Ⓚ

© 2006 Kaplan, Inc.

MATHEMATICS TEST
60 Minutes—60 Questions

DIRECTIONS: Solve each problem, choose the correct answer, and then fill in the corresponding oval on your answer document.

Do not linger over problems that take too much time. Solve as many as you can; then return to the others in the time you have left for this test.

You are permitted to use a calculator on this test. You may use your calculator for any problems you choose, but some of the problems may best be done without using a calculator.

Note: Unless otherwise stated, all of the following should be assumed.

1. Illustrations are NOT necessarily drawn to scale.
2. Geometric figures lie in a plane.
3. The word *line* indicates a straight line.
4. The word *average* indicates arithmetic mean.

---

1. Ten students are receiving honors credit for taking Mr. Friedman's class. This number is exactly 20% of the total number of students in the class. How many students are in Mr. Friedman's class?

    A. 12
    B. 15
    C. 18
    D. 20
    E. 50

DO YOUR FIGURING HERE.

2. In the figure below, points *A*, *B*, and *C* are on a straight line. What is the measure of angle *DBE*?

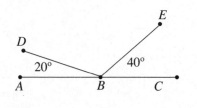

    F. 60°
    G. 80°
    H. 100°
    J. 120°
    K. 140°

GO ON TO THE NEXT PAGE.

© 2006 Kaplan, Inc.

3. What is the fifth term of the arithmetic sequence 7, 4, 1, … ?

- A. ⁻5
- B ⁻2
- C. 1
- D. 4
- E. 14

4. Which of the following values of c is the solution to the proportion $\frac{20}{8} = \frac{c}{10}$?

- F. 4
- G. 16
- H. 18
- J. 22
- K. 25

5. If G, H, and K lie on the same line, and $\overline{GK} \approx \overline{HK}$, then which of the following must be true?

- A. G is the midpoint of $\overline{HK}$.
- B. H is the midpoint of $\overline{GK}$.
- C. K is the midpoint of $\overline{GH}$.
- D. G is the midpoint of $\overline{KH}$.
- E. K is the midpoint of $\overline{KG}$.

6. Four pieces of yarn, each 1.2 meters long, were cut from the end of a ball of yarn that was 50 meters long before the four pieces were cut. How many meters of yarn are left?

- F. 45.2
- G. 45.8
- H. 46.8
- J. 47.2
- K. 47.8

GO ON TO THE NEXT PAGE.

ACT ADVANTAGE
MATHEMATICS

© 2006 Kaplan, Inc.

7. If $x = {}^-2$, then $14 - 3(x + 3) = $?

DO YOUR FIGURING HERE.

    A. $^-1$

    B. 11

    C. 14

    D. 17

    E. 29

8. $^-|^-6| - (^-6) = $?

    F. $^-36$

    G. $^-12$

    H. 0

    J. 12

    K. 36

9. A car dealership expects an increase of 15% in its current annual sales of 3,200 cars. What will be its new annual sales?

    A. 3,215

    B. 3,248

    C. 3,680

    D. 4,700

    E. 4,800

10. If $x^4 = 90$ and x is a real number, then x lies between which two consecutive integers?

    F. 2 and 3

    G. 3 and 4

    H. 4 and 5

    J. 5 and 6

    K. 6 and 7

A

GO ON TO THE NEXT PAGE.

© 2006 Kaplan, Inc.

11. If $47 - x = 188$, then $x = ?$

    A. $^-235$

    B. $^-141$

    C.   4

    D.   141

    E.   235

DO YOUR FIGURING HERE.

12. Group A needs 8 more hours than Group B in order to paint a wall. Group B needs twice as long as Group C to complete the same task. If Group C took h hours to paint a wall in Katy's house, how many hours would the task have taken Group A?

    F. $10h$

    G. $16h$

    H. $10 + h$

    J. $2(8 + h)$

    K. $8 + 2h$

13. In the standard (x, y) coordinate plane, three corners of a rectangle are located at $(2, ^-2)$, $(^-5, ^-2)$, and $(2, ^-5)$. What is the location of the rectangle's fourth corner?

    A. $(2, 5)$

    B. $(^-2, 5)$

    C. $(^-2, 2)$

    D. $(^-2, ^-5)$

    E. $(^-5, ^-5)$

A

14. Which of the following is another way to write $5a - 5b + 3a$?

    F. $5(a - b + 3)$

    G. $(a - b)(5 + 3a)$

    H. $a(8 - 5b)$

    J. $8a - 5b$

    K. $2a - 5b$

GO ON TO THE NEXT PAGE.

© 2006 Kaplan, Inc.

15. What is the measure of angle *FEG* in the parallelogram below?

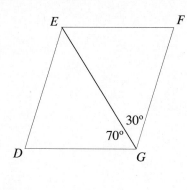

   A. 30°
   B. 40°
   C. 50°
   D. 60°
   E. 70°

16. What is the slope of any line parallel to the line $4x + 3y = 9$?

   F. $^-4$

   G. $-\dfrac{4}{3}$

   H. $\dfrac{4}{9}$

   J. 4

   K. 9

17. If $x > 0$ and $3x^2 - 7x - 20 = 0$, then $x = ?$

   A. $\dfrac{5}{3}$
   B. 3
   C. 4
   D. 7
   E. 20

GO ON TO THE NEXT PAGE.

© 2006 Kaplan, Inc.

18. The lengths of the sides of a triangle are 2, 5, and 8 centimeters. How many centimeters long is the shortest side of a similar triangle that has a perimeter of 30 centimeters?

F. 4

G. 7

H. 10

J. 15

K. 16

19. A shirt that normally sells for $24.60 is on sale for 15% off. How much does it cost during the sale, to the nearest dollar?

A. $ 4

B. $10

C. $20

D. $21

E. $29

20. Which of the following is a factored form of $3xy^4 + 3x^4y$?

F. $3x^4y^4(y + x)$

G. $3xy(y^3 + x^3)$

H. $6xy(y^3 + x^3)$

J. $3x^4y^4$

K. $6x^5y^5$

21. If $x - 2y = 0$ and $3x + y = 7$, what is the value of x?

A. $^-1$

B. 0

C. 1

D. 2

E. 3

DO YOUR FIGURING HERE.

GO ON TO THE NEXT PAGE.

© 2006 Kaplan, Inc.

22. There are three feet in a yard. If 2.5 yards of fabric at Fabric World cost $4.50, what is the fabric's price per foot?

F. $ 0.60

G. $ 0.90

H. $ 1.50

J. $ 1.80

K. $11.25

DO YOUR FIGURING HERE.

23. The figure below shows a square overlapping a rectangle. One vertex of the rectangle is in the center of the square. What is the area, in square units, of the shaded region?

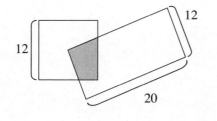

A. 9

B. 18

C. 36

D. 72

E. 144

GO ON TO THE NEXT PAGE.

© 2006 Kaplan, Inc.

24. Caleb's floor has the dimensions shown below. How many square feet of tile will Caleb need in order to cover his entire floor? (Note: All angles are right angles.)

DO YOUR FIGURING HERE.

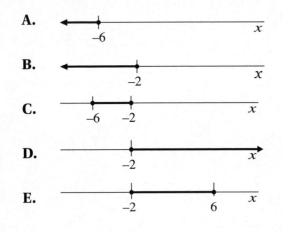

14 feet

20 feet

8 feet

22 feet

F.   64
G.   96
H.   160
J.   344
K.   484

25. Which of the following is the graph of the inequality $x - 2 \leq {}^{-}4$?

A.

B.

C.

D.

E.

GO ON TO THE NEXT PAGE.

© 2006 Kaplan, Inc.

26. Betty sells jewelry. She earns $7h + 0.04s$ dollars, where h is the number of hours worked, and s is the total price of the jewelry sold. What would she earn for working 15 hours and selling $120.50 worth of jewelry?

F. $109.82

G. $153.20

H. $226.10

J. $231.50

K. $848.32

DO YOUR FIGURING HERE.

27. Which of the following is less than $\frac{3}{5}$?

A. $\frac{4}{6}$

B. $\frac{8}{13}$

C. $\frac{6}{10}$

D. $\frac{7}{11}$

E. $\frac{4}{7}$

28. What is the area, in square feet, of a right triangle with sides of lengths 7 feet, 24 feet, and 25 feet?

F. 56

G. 84

H. $87\frac{1}{2}$

J. 168

K. 300

GO ON TO THE NEXT PAGE.

© 2006 Kaplan, Inc.

29. When Mike's graduating class is arranged in rows of 6 people each, the last row is one person short. When it is arranged in rows of 7, the last row is still one person short. When the class is arranged in rows of 8, the last row is *still* one person short. What is the fewest possible number of people in Mike's graduating class?

DO YOUR FIGURING HERE.

- **A.** 23
- **B.** 41
- **C.** 71
- **D.** 167
- **E.** 335

30. The lengths of two sides of a triangle are 3.5 inches and 6 inches. Which of the following CANNOT be the length, in inches, of the third side?

- **F.** 2
- **G.** 3
- **H.** 4
- **J.** 5
- **K.** 6

31. For all $b > 0$, $\dfrac{4}{5} + \dfrac{1}{b} = ?$

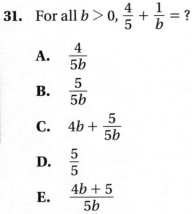

- **A.** $\dfrac{4}{5b}$
- **B.** $\dfrac{5}{5b}$
- **C.** $4b + \dfrac{5}{5b}$
- **D.** $\dfrac{5}{5}$
- **E.** $\dfrac{4b + 5}{5b}$

GO ON TO THE NEXT PAGE.

ACT ADVANTAGE
MATHEMATICS

© 2006 Kaplan, Inc.

32. How long is side $\overline{EF}$ in the right triangle below?

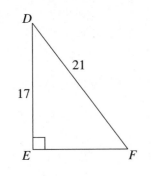

F. $\sqrt{21^2 - 17^2}$

G. $\sqrt{21^2 + 17^2}$

H. $21^2 - 17^2$

J. $21^2 + 17^2$

K. $\sqrt{21 - 17}$

33. The measure of each side of square *JKLM* is *b* inches. If its length is increased by 2 inches and its width is increased by 3 inches, a rectangle will be formed. What will be the area, in square inches, of the new rectangle?

A. $2b + 5$

B. $4b + 10$

C. $b^2 + 6$

D. $b^2 + 5b + 5$

E. $b^2 + 5b + 6$

34. If $\sin \beta = \dfrac{8}{17}$, and $\cos \beta = \dfrac{15}{17}$, what is the value of $\tan \beta$?

F. $\dfrac{7}{17}$

G. $\dfrac{8}{15}$

H. $\dfrac{23}{17}$

J. $\dfrac{15}{8}$

K. $\dfrac{120}{17}$

GO ON TO THE NEXT PAGE.

© 2006 Kaplan, Inc.

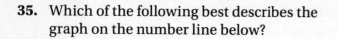

35. Which of the following best describes the graph on the number line below?

DO YOUR FIGURING HERE.

$$-3 \quad -2 \quad -1 \quad 0 \quad 1 \quad x$$

A. $^-|x| = ^-2$

B. $^-|x| < 0.5$

C. $^-3 < x < ^-1$

D. $^-1.5 < x < ^-2.5$

E. $^-1.5 > x > ^-2.5$

36. Carolyn's basketball team made 1-point, 2-point, and 3-point shots during their last game. Twenty percent of their baskets were worth 1 point, 70% of their baskets were worth 2 points, and 10% of their baskets were worth 3 points. To the nearest tenth of a point, what was the average point value of the shots made by Carolyn's basketball team?

F. 1.4

G. 1.7

H. 1.8

J. 1.9

K. 2.0

37. What is the largest possible product of two odd integers whose sum is 42?

A. 117

B. 185

C. 259

D. 377

E. 441

GO ON TO THE NEXT PAGE.

© 2006 Kaplan, Inc.

38. In the triangle below, if $\overline{CD}$ is 3 centimeters long, what is the length of $\overline{CE}$?

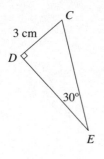

F. 3 cm

G. $3\sqrt{2}$ cm

H. $3\sqrt{3}$ cm

J. 6 cm

K. 9 cm

39. In the (x, y) coordinate plane, what is the y-intercept of the line $12x - 3y = 12$?

A. $^-4$

B. $^-3$

C. 0

D. 4

E. 12

40. In the figure below, lines l and m are parallel, and lines n and p are parallel. What is the value of x?

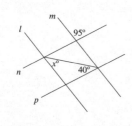

F. 40

G. 45

H. 50

J. 70

K. 85

GO ON TO THE NEXT PAGE.

© 2006 Kaplan, Inc.

41. Which of the points graphed on the number line below is closest to the value of e? (Note: $e \approx 2.718281828$)

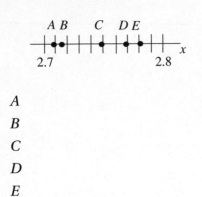

A. A

B. B

C. C

D. D

E. E

DO YOUR FIGURING HERE.

42. For which value of a does the following system of equations have no solution?

$$^-x + 6y = 7$$
$$^-5x + 10ay = 32$$

F. $\dfrac{5}{3}$

G. 3

H. 6

J. 30

K. 60

43. The expression $(360 - x)°$ is the measure of a nonzero obtuse angle if and only if:

A. $0 < x < 90$

B. $0 < x < 180$

C. $180 < x < 270$

D. $180 < x < 360$

E. $270 < x < 360$

GO ON TO THE NEXT PAGE.

ACT ADVANTAGE
MATHEMATICS

© 2006 Kaplan, Inc.

44. If $p - q = {}^-4$ and $p + q = {}^-3$,
then $p^2 - q^2 = $?

F. $^-12$

G. $^-7$

H.   7

J.   12

K.   25

45. The lengths of the sides of a triangle are 6
meters, 8 meters, and 10 meters. What is the
angle formed by the intersection of the two
shortest sides?

A.   30°

B.   45°

C.   60°

D.   90°

E.   135°

46. In the standard (x, y) coordinate plane, if
the x-coordinate of each point on a line is 9
more than three times the y-coordinate, the
slope of the line is:

F. $^-9$

G. $^-3$

H. $\dfrac{1}{3}$

J.   3

K.   9

A

GO ON TO THE NEXT PAGE.

© 2006 Kaplan, Inc.

47. A tree is growing at the edge of a cliff, as shown below. From the tree, the angle between the base of the cliff and the base of a nearby house is 62°. If the distance between the base of the cliff and the base of the house is 500 feet, how many feet tall is the cliff?

DO YOUR FIGURING HERE.

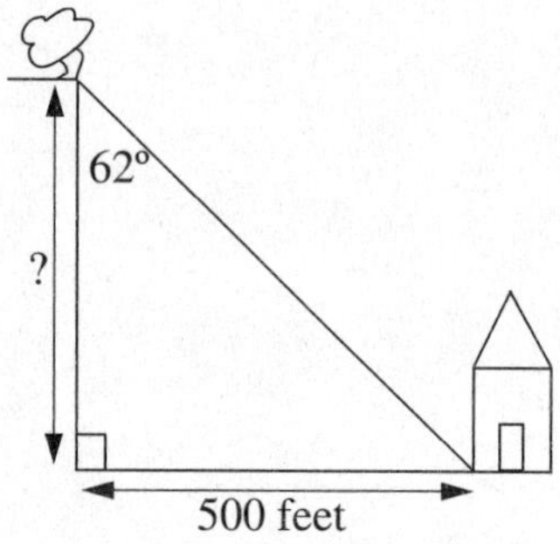

A. $500 \cos 62°$

B. $500 \tan 62°$

C. $\dfrac{500}{\sin 62°}$

D. $\dfrac{500}{\cos 62°}$

E. $\dfrac{500}{\tan 62°}$

48. Two numbers have a greatest common factor of 9 and a least common multiple of 54. Which of the following is the pair of numbers?

F.   9 and 18

G.   9 and 27

H.   18 and 27

J.   18 and 54

K.   27 and 54

GO ON TO THE NEXT PAGE.

© 2006 Kaplan, Inc.

49. Five functions are listed below. If k is a real number less than 1, and $a(x) = 5^x$, which of these functions yields the greatest value of $a(b(x))$ for all $x > 2$?

DO YOUR FIGURING HERE.

A. $b(x) = \dfrac{k}{x}$

B. $b(x) = \dfrac{x}{k}$

C. $b(x) = kx$

D. $b(x) = x^k$

E. $b(x) = \sqrt[k]{x}$

50. Line segments $\overline{WX}$, $\overline{XY}$, and $\overline{YZ}$ form the rectangular box shown below, and have lengths of 12 centimeters, 5 centimeters, and 13 centimeters, respectively. What is the cosine of $\angle ZWY$?

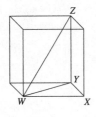

F. $\dfrac{5}{13}$

G. $\dfrac{\sqrt{2}}{2}$

H. $\dfrac{12}{13}$

J. 1

K. $\dfrac{13\sqrt{2}}{12}$

51. A certain circle has an area of 4π square centimeters. What is the length, in centimeters, of its radius?

A. $\dfrac{1}{4}$

B. 2

C. 4

D. 2π

E. 4π

GO ON TO THE NEXT PAGE.

© 2006 Kaplan, Inc.

52. The equation of line *l* below is
$y = mx + b$. Which of the following
could be the equation of line *q*?

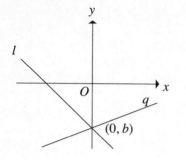

F. $y = \frac{1}{2}mx$

G. $y = \frac{1}{2}mx - b$

H. $y = \frac{1}{2}mx + b$

J. $y = -\frac{1}{2}mx - b$

K. $y = -\frac{1}{2}mx + b$

53. The equation $x^2 - 6x + k = 0$ has exactly
one solution for *x*. What is the value of *k*?

A. 0

B. 3

C. 6

D. 9

E. 12

54. What is the slope of the line that passes
through the origin and $(\frac{1}{3}, \frac{3}{4})$?

F. $\frac{1}{4}$

G. $\frac{1}{3}$

H. $\frac{5}{12}$

J. $\frac{3}{4}$

K. $\frac{9}{4}$

GO ON TO THE NEXT PAGE.

ACT ADVANTAGE
MATHEMATICS

© 2006 Kaplan, Inc.

DO YOUR FIGURING HERE.

55. If R, S, and T are real numbers, and $RST = 2$, which of the following *must* be true?

A. $RT = \dfrac{2}{S}$

B. R, S, and T are all positive.

C. $R = 2$, $S = 2$, or $T = 2$

D. $R = 0$, $S = 0$, or $T = 0$

E. $R > 2$, $S > 2$, or $T > 2$

56. A square has sides of length $(w + 5)$ units. If a rectangle with a length of $(w + 2)$ units and a width of $(w - 3)$ units is removed from the interior of the square, which of the following is the remaining area of the square?

F. 31 square units

G. $9w + 19$ square units

H. $11w + 31$ square units

J. $w^2 + 10w + 25$ square units

K. $2w^2 + 9w + 19$ square units

57. What is the smallest positive value for θ where $\sin 2\theta$ reaches its minimum value?

A. $\dfrac{\pi}{4}$

B. $\dfrac{\pi}{2}$

C. $\dfrac{3\pi}{4}$

D. π

E. $\dfrac{3\pi}{2}$

GO ON TO THE NEXT PAGE.

© 2006 Kaplan, Inc.

58. In the standard (x, y) coordinate plane, if the distance between the points $(r, 6)$ and $(10, r)$ is 4 units, which of the following is the value of r?

DO YOUR FIGURING HERE.

- **F.** 3
- **G.** 4
- **H.** 7
- **J.** 8
- **K.** 10

59. Calleigh put 5 nickels into an empty hat. She wants to add enough pennies so that the probability of drawing a nickel at random from the hat is $\frac{1}{6}$. How many pennies should she put into the hat?

- **A.** 1
- **B.** 5
- **C.** 10
- **D.** 25
- **E.** 30

60. How many different integer values of x satisfy the inequality $\frac{1}{5} < \frac{3}{x} < \frac{1}{3}$?

- **F.** 1
- **G.** 2
- **H.** 3
- **J.** 4
- **K.** 5

A

END OF TEST.

STOP! DO NOT TURN THE PAGE UNTIL TOLD TO DO SO.

© 2006 Kaplan, Inc.

The Landscape of the Test

Thinking KAP

You're going to play a new video game with your friends, and you want to score as many points as possible. Before you begin, you have a chance to ask someone who has played the game for advice. What kinds of questions will you ask to ensure that you score well?

B

© 2006 Kaplan, Inc.

Strategy Instruction

Know Where the Points Are

Each year, college admissions officers use ACT scores to compare applicants. The ACT Test that is administered in any given year is never used again, so students cannot copy the answers and provide them to students who take the test the following year. How, then, can admissions officers compare their applicants' scores if the tests they took were not identical?

The answer is simple. The ACT is written to follow clearly defined specifications that are used each time a new test is created. Different ACTs test the same content with the same kinds of problems, but no two tests are identical, or ever contain the exact same problems.

Because the ACT does not change much between administrations, you can anticipate topics that will be covered on the test and study them ahead of time so that you are well prepared.

keep in mind

You will also receive three sub-scores on the ACT Mathematics Test: one for pre-algebra/ elementary algebra, one for intermediate algebra/ coordinate geometry, and one for plane geometry/trigonometry.

Topic	Approximate # of Questions	Approximate Point Value
Pre-Algebra	14	8.3
Algebra	19	11.5
Coordinate Geometry	9	5.4
Plane Geometry	14	8.3
Trigonometry	4	2.5
Total	60	36

B

© 2006 Kaplan, Inc.

Know Which Tools You Can Use on the Test

Calculator

You may use a calculator, including a scientific or graphing calculator, on the ACT Mathematics Test (but not on any of the other ACT tests). Keep the following points in mind:

1) To ensure that your calculator will be permitted, go to www.actstudent.org before Test Day and check the regularly updated lists, or call 1-800-498-6481 for a recorded message.

2) Bring spare batteries and/or a spare calculator.

3) Remove all power cords.

4) Turn off the sound on your calculator.

5) Remove any paper tape.

6) Cover infrared ports with heavy tape.

keep in mind

When guessing, try to eliminate answers. Even if you don't know the correct answer, you can improve your odds of success with each choice you rule out.

Some Types of Prohibited Calculators
Calculators with algebra systems
Calculators with letter keys
Cell-phone calculators
Electronic writing pads
Handheld computers
Notebook computers
Pocket organizers

Guessing

There is no penalty for incorrect answers on the Mathematics ACT Test. Therefore, you should answer every question, even if you have to guess.

B

© 2006 Kaplan, Inc.

Know How to Pace Yourself

The ACT Mathematics Test is 60 minutes long. That's only 1 minute per question! To maximize your performance, follow these tips:

Answer the Easy Questions First

All of the questions on the ACT Mathematics Test are worth the same number of points. If you get bogged down on one or two difficult questions, you might run out of time before you have a chance to answer some questions that may be relatively easy for you.

Keep Track of Time

Bring a watch in case your seat doesn't allow you a clear view of a clock. Check the time after every few questions. Be sure to pace yourself and spend an average of less than 1 minute per question.

Leave Time to Check Your Work

Everyone makes simple mistakes once in a while. Plan ahead so you will have enough time to check your calculations.

keep in mind

Place a "?" in the margin next to any questions you skip so you won't forget to return to them later.

© 2006 Kaplan, Inc.

Learn From Your Mistakes

Practicing to take the ACT is important, but practice is not enough. You need to review your progress and build on the skills that you already have. Throughout this program, pay particular attention to the questions that you answer incorrectly. You will likely learn more from your mistakes than from your successes.

Start by listing the questions you answered incorrectly on the Practice Test.

Topic	Problems	Correct	Incorrect
Pre-Algebra	1, 3, 6, 8, 9, 19, 22, 25, 27, 29, 36, 37, 41, 48, 59		
Algebra	4, 7, 10, 11, 12, 14, 17, 20, 21, 26, 31, 35, 42, 44, 49, 53, 55, 60		
Plane Geometry	2, 15, 18, 23, 24, 28, 30, 32, 33, 38, 40, 43, 45, 50, 51, 56		
Coordinate Geometry	5, 13, 16, 39, 46, 52, 54, 58		
Trigonometry	34, 47, 57		

keep in mind

Flag pages containing material that is confusing to you. Then be sure to review them before Test Day.

B

© 2006 Kaplan, Inc.

Guided Practice

Look at these problems that you solved in the Practice Test.

1. When Mike's graduating class is arranged in rows of 6 people each, the last row is one person short. When it is arranged in rows of 7, the last row is still one person short. When the class is arranged in rows of 8, the last row is *still* one person short. What is the fewest possible number of people in Mike's graduating class?

    A. 23
    B. 41
    C. 71
    D. 167
    E. 335

Look back at your work on the Practice Test.

-   What kind of math is needed for this problem?

    _____

-   What is the important information in this problem?

    _____

-   What did you do to solve this problem on the test?

    _____

-   Did you solve this problem correctly, or can you find a mistake in your work?

    _____

© 2006 Kaplan, Inc.

2. A square has sides of length $(w + 5)$ units. If a rectangle with a length of $(w + 2)$ units and a width of $(w − 3)$ units is removed from the interior of the square, which of the following is the remaining area of the square?

    F.   31 square units

    G. $9w + 19$ square units

    H. $11w + 31$ square units

    J. $w^2 + 10w + 25$ square units

    K. $2w^2 + 9w + 19$ square units

Look back at your work on the Practice Test.

- What kind of math is needed for this problem?

    _____

- What is the important information in this problem?

    _____

- What did you do to solve this problem on the test?

    _____

- Did you solve this problem correctly, or can you find a mistake in your work?

    _____

B

© 2006 Kaplan, Inc.

Shared Practice

Solve each problem below. Use the hints provided to help you.

1. Line m is perpendicular to the line that passes through the points (5, 6) and (6, 10). What is the slope of line m?

   A. $^-4$

   B. $-\dfrac{1}{4}$

   C. $\dfrac{1}{4}$

   D. 4

   E. 8

   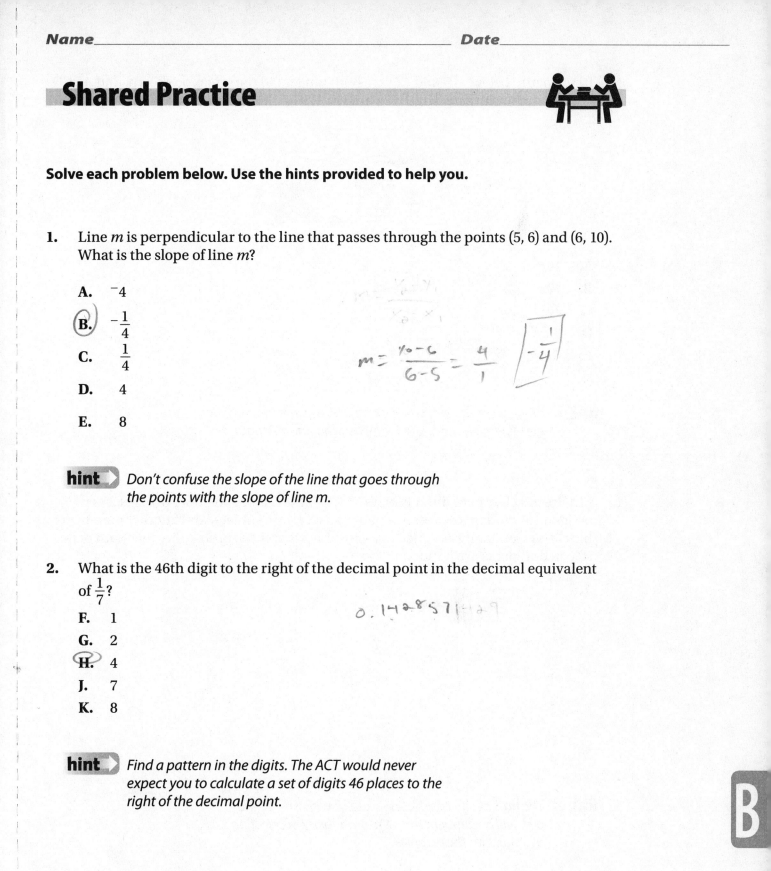

   hint ▶ *Don't confuse the slope of the line that goes through the points with the slope of line m.*

2. What is the 46th digit to the right of the decimal point in the decimal equivalent of $\dfrac{1}{7}$?

   F. 1

   G. 2

   H. 4

   J. 7

   K. 8

   hint ▶ *Find a pattern in the digits. The ACT would never expect you to calculate a set of digits 46 places to the right of the decimal point.*

© 2006 Kaplan, Inc.

3. In the figure below, $\overline{AB}$ and $\overline{CD}$ are both tangent to the circle, as shown, and *ABCD* is a rectangle with side lengths 2*x* and 5*x*. What is the area of the shaded region?

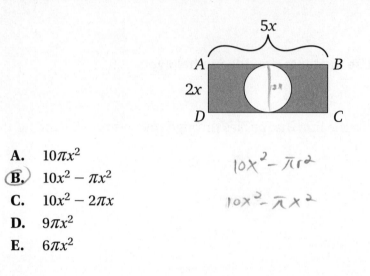

A. $10\pi x^2$

B. $10x^2 - \pi x^2$

C. $10x^2 - 2\pi x$

D. $9\pi x^2$

E. $6\pi x^2$

$10x^2 - \pi r^2$

$10x^2 - \pi x^2$

 hint ▷ *When working with complex figures, use information pertaining to one figure for measuring another figure.*

4. Mr. Breen's class contains 5 juniors and 5 seniors. If one member of his class is assigned at random to present a paper on a certain subject, and another member of his class is randomly assigned to assist him, what is the probability that both of the selected students are juniors?

F. $\dfrac{1}{10}$

G. $\dfrac{1}{5}$

H. $\dfrac{2}{9}$

J. $\dfrac{2}{5}$

K. $\dfrac{1}{2}$

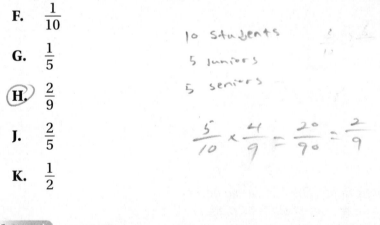

10 students

5 juniors

5 seniors

$\dfrac{5}{10} \times \dfrac{4}{9} = \dfrac{20}{90} = \dfrac{2}{9}$

hint ▷ *The probability of selecting a junior with the second pick will be different from the probability of selecting a junior with the first pick.*

© 2006 Kaplan, Inc.

5. In the right triangle below, sin θ = ?

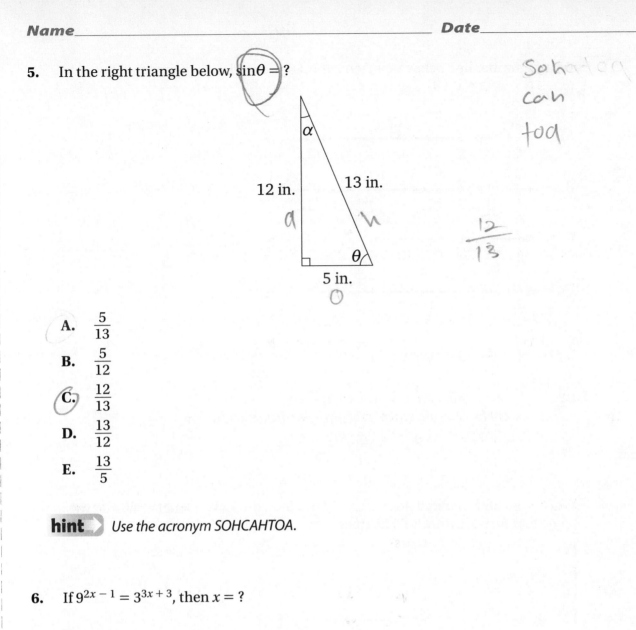

12 in.

13 in.

5 in.

α

θ

handwritten: Sohatoa
can
toa

$\frac{12}{13}$

A. $\frac{5}{13}$

B. $\frac{5}{12}$

C. $\frac{12}{13}$

D. $\frac{13}{12}$

E. $\frac{13}{5}$

hint ▶ *Use the acronym SOHCAHTOA.*

6. If $9^{2x-1} = 3^{3x+3}$, then $x = $?

F. $^-4$

G. $\frac{-7}{4}$

H. $\frac{-10}{7}$

K. 2

J. 5

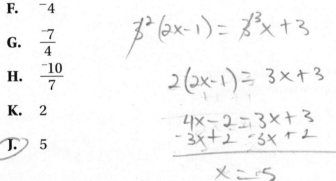

hint ▶ *Factor one side until the bases are the same, then set the exponents equal and solve.*

© 2006 Kaplan, Inc.

B

7. Which number line below represents the solutions for *x* in the inequality
$5x - 2(1 - x) \geq 4(x + 1)$?

handwritten:
$5x - 2 + 2x \geq 4x + 4$

$7x - 2 \geq 4x + 4$
$-4x + 2 \quad -4x + 2$
_____
$\dfrac{3x}{3} \geq \dfrac{6}{3}$

$x \geq 2$

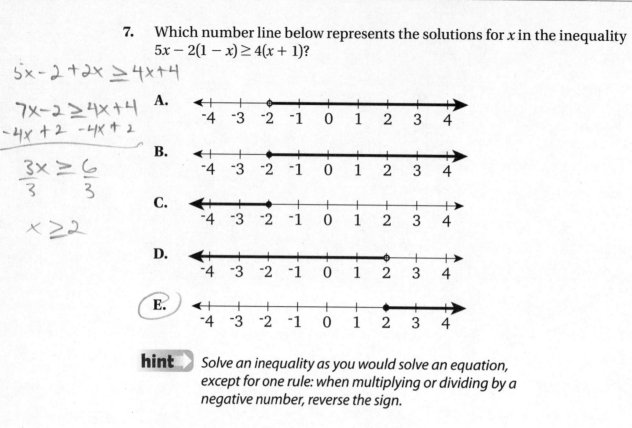

A.

B.

C.

D.

E.

hint ▷ *Solve an inequality as you would solve an equation, except for one rule: when multiplying or dividing by a negative number, reverse the sign.*

8. What is the area of parallelogram *ABCD* if all 4 sides have a length of 6 and the angle measure of ∠*BAD* is 60 degrees?

F. 18
G. 24
H. $18\sqrt{3}$
J. 36
K. $24\sqrt{3}$

handwritten:
$long = x\sqrt{3}$
$short = x$
$hyp = 2x$

$6 \times 3\sqrt{3} =$
$\boxed{18\sqrt{3}}$

$6 = 2x$

$3\sqrt{3}$ 6
3

hint ▷ *If the question doesn't supply a diagram, draw your own so that you can visualize what is going on.*

© 2006 Kaplan, Inc.

B

KAP Wrap

Write a letter to your teacher, telling him or her your thoughts about the ACT. In your letter, make sure to say:

- what kind of math you think you are strong in.

- what kind of math you think you need to work on.

- what you think you can do to get ready for the test.

- how you think your teacher can help you get ready for the test.

B

© 2006 Kaplan, Inc.

lesson C
Strategies for Success

ReKAP

Review the strategies from Lessons A and B. Then fill in the blanks with what you have learned.

1. I will have _____ minutes to answer _____ questions on the ACT Mathematics Test.

2. I should write all of my answers in the _____ provided.

3. Since all of the problems on the ACT Mathematics Test are worth _____ points, I should answer every question, even if I have to make a few guesses.

4. When solving a problem containing a diagram, I cannot assume that it is drawn _____ unless told so.

5. I should study hard so I will be well-prepared to take the ACT on _____.

C

© 2006 Kaplan, Inc.

Smart Studying

You can learn a lot in a short amount of time if you follow some simple guidelines for smart studying. Follow these simple steps:

Plan Your Personal Study Time

Choose a distraction-free location that is comfortable and convenient, such as a space in your home where other family members won't engage you, or a library. Set aside regular times during the week to review the concepts that you learn in this course.

Make Flash Cards

Use index cards to make flash cards for terms or concepts that you find difficult to remember. Write the term or concept on one side, and the definition and an example of how to use it on the other side. Bind them with a rubber band and keep them handy so you can review them and test yourself on a regular basis.

Form a Study Group

Invite three or four of your classmates to create a study group with you. Find a quiet location and decide how often you want to meet. Select the day(s) and time(s) your group will get together during the week, and study every week until Test Day.

 *Try It Out!*

What is your smart studying plan?

- Which days of the week are most convenient for you to study? _____

  _____

- Select a location where you can study. _____

- Name two math topics that you find confusing. _____

- Find three or four classmates and form a study group. Write the names of the study group members.

  _____

- Where and when will you meet each week? _____

The better you make use of the resources available to you, the more likely you will do your best on Test Day.

© 2006 Kaplan, Inc.

Smart Test-Taking

Know how to put your hard-earned knowledge to work on Test Day. Follow these smart test-taking tips:

Take What You Need to the Test Center

Make sure you bring the items below with you on Test Day:

- your admission ticket
- acceptable identification (see www.actstudent.org to confirm the latest rules on ID)
- three sharpened (NOT mechanical) #2 pencils with good erasers
- an approved calculator that you've practiced using before the test
- a non-alarm watch
- glasses or contact lenses (if you use them)

You might want to also bring a bottle of water, facial tissues, and a snack to eat outside the test room during breaks.

keep in mind

Knowing what not to do when taking a test is often as helpful as knowing what to do.

Arrive on Time

The rules of the ACT state that test takers are not to be admitted after test booklets have been distributed. Make sure you know where to take the test before Test Day, and how to get there.

Clearly Mark Problems That You Want to Complete or Check Later

Write a large question mark to the left of each problem that you need to complete or double check later. Do your work to the right of the problem.

Don't Lose Track of Time

After every 10 problems you answer, check the time to make sure you are working at an appropriate pace. If you get stuck on a problem, mark it and return to it later.

Don't Mark the Correct Answer in the Wrong Place

Take a quick glance at your Answer Sheet after every 10 questions to be sure that the next question in the test booklet corresponds to the next answer space on your Answer Sheet.

© 2006 Kaplan, Inc.

MATHEMATICS TEST

15 Minutes—15 Questions

DIRECTIONS: Solve each problem, choose the correct answer, and then fill in the corresponding oval on your answer document.

Do not linger over problems that take too much time. Solve as many as you can; then return to the others in the time you have left for this test.

You are permitted to use a calculator on this test. You may use your calculator for any problems you choose, but some of the problems may best be done without using a calculator.

Note: Unless otherwise stated, all of the following should be assumed.

1. Illustrations are NOT necessarily drawn to scale.
2. Geometric figures lie in a plane.
3. The word *line* indicates a straight line.
4. The word *average* indicates arithmetic mean.

---

1. How many inches tall would a cylinder with a diameter of 4 inches have to be to have the same volume as a cylinder with a diameter of 6 inches and a height of 4 inches?

   A. 3
   B. 6
   C. 9
   D. 12
   E. 18

DO YOUR FIGURING HERE.

$$V = \pi r^2 \cdot h$$

$$\pi 2^2 \times h = \pi 3^2 \times 4$$

$$\pi 4h = \pi 9 \times 4$$

$$\frac{4h}{4} = \frac{36}{4}$$

$$h = 9$$

2. What is the value of x if:

   $$\frac{1}{x} + \frac{2}{x} + \frac{3}{x} + \frac{4}{x} = 5?$$

   F. $\dfrac{1}{2}$
   G. 2
   H. 4
   J. $12\dfrac{1}{2}$
   K. 50

GO ON TO THE NEXT PAGE.

© 2006 Kaplan, Inc.

3. The two overlapping circles below form three regions, as shown:

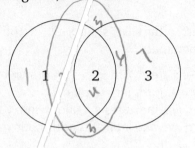

What is the maximum number of regions that can be formed by three overlapping circles?

A. 5
B.
C. 7
D. 8
F. 9

4. What is the sum of $\frac{3}{16}$ and 0.175?

F. 0.3165
G. 0.3500
H. 0.3625
J. 0.3750
K. 0.3875

5. For all x, $3x^2 \times 5x^3 = ?$

A. $8x^5$
B. $8x^6$
C. $15x^5$
D. $15x^6$
E. $15x^8$

GO ON TO THE NEXT PAGE.

© 2006 Kaplan, Inc.

6. In the figure below, $\overline{QS}$ and $\overline{PT}$ are parallel, and the lengths of $\overline{QR}$ and $\overline{PQ}$, in units, are as marked. If the perimeter of $\triangle QRS$ is 11 units, what is the perimeter of $\triangle PRT$?

DO YOUR FIGURING HERE.

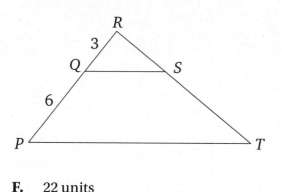

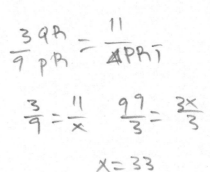

F.   22 units

G.   33 units

H.   66 units

J.   88 units

K.   99 units

7. Nine less than a number, c, is equal to d, and d less than twice c is 20. Which two equations could be used to determine the values of c and d?

A. $d - 9 = c$
     $d - 2c = 20$

B. $c - 9 = d$
     $2c - d = 20$

C. $c - 9 = d$
     $d - 2c = 20$

D. $9 - c = d$
     $2c - d = 20$

E. $9 - c = d$
     $2cd = 20$

GO ON TO THE NEXT PAGE.

© 2006 Kaplan, Inc.

8. Which of the following equations describes the parabola graphed below?

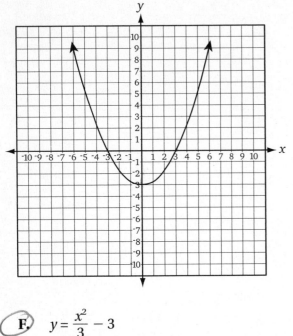

$y =$?² − 3

F. $y = \dfrac{x^2}{3} - 3$

G. $y = x^2 - \dfrac{3}{3}$

H. $y = \dfrac{x^2}{3} + 3$

J. $y = x^2 + \dfrac{3}{3}$

K. $y = 3x^2 - 3$

9. One number is 5 times another number, and their sum is ⁻60. What is the lesser of the two numbers?

A. ⁻50

B. ⁻48

C. ⁻12

D. ⁻10

E. ⁻5

GO ON TO THE NEXT PAGE.

10. Compared to the graph of $y = \cos\theta$, the graph of $y = 2\cos\theta$ has:

    F. twice the period and the same amplitude.

    G. half the period and the same amplitude.

    H. twice the period and half the amplitude.

    J. half the amplitude and the same period.

    K. twice the amplitude and the same period.

11. If $^-3$ is a solution for the equation $x^2 + kx - 15 = 0$, what is the value of k?

    A. $^-5$

    B. $^-2$

    C. 2

    D. 5

    E. cannot be determined from the information given

DO YOUR FIGURING HERE.

$9 - 3x - 15 = 0$

$-6 - 3x = 0$

$+6 \qquad +6$

$\dfrac{-3x}{-3} = \dfrac{6}{-3}$

$x = -2$

GO ON TO THE NEXT PAGE.

© 2006 Kaplan, Inc.

12. In the standard (x, y) coordinate plane shown below, Points A and B lie on line m, and Point C lies below it. The coordinates of Points A, B, and C are $(0, 5)$, $(5, 5)$, and $(3, 3)$, respectively. What is the shortest possible distance from Point C to a point on line m?

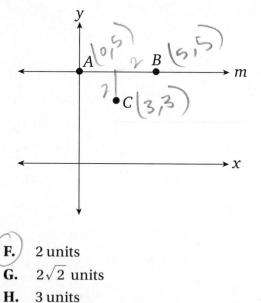

F. 2 units

G. $2\sqrt{2}$ units

H. 3 units

J. $\sqrt{13}$ units

K. 5 units

13. In the standard (x, y) coordinate plane, the coordinates of Points P and Q are $(2, 3)$ and $(12, {}^-15)$, respectively. What are the coordinates of the midpoint of $\overline{PQ}$?

A. $(6, {}^-12)$

B. $(6, {}^-9)$

C. $(6, {}^-6)$

D. $(7, {}^-9)$

E. $(7, {}^-6)$

midpoint $\left(\dfrac{x_1 + x_2}{2}, \dfrac{y_1 + y_2}{2} \right)$

GO ON TO THE NEXT PAGE.

© 2006 Kaplan, Inc.

14. What is the perimeter of a 30°60°90° triangle in which the length of the longer leg is 12 inches?

DO YOUR FIGURING HERE.

F. $(6\sqrt{3} + 12)$ inches

G. $(4\sqrt{3} + 18)$ inches

H. $(8\sqrt{3} + 18)$ inches

J. $(12\sqrt{3} + 12)$ inches

K. $(12\sqrt{3} + 18)$ inches

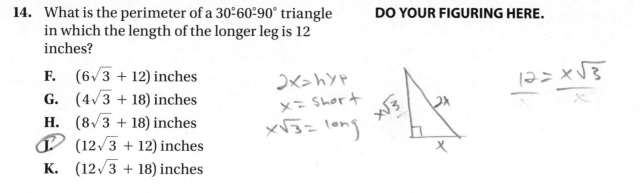

15. In a group of 25 students, 16 are female. What percentage of the group is female?

A. 16%

B. 40%

C. 60%

D. 64%

E. 75%

END OF TEST. C

STOP! DO NOT TURN THE PAGE UNTIL TOLD TO DO SO.

© 2006 Kaplan, Inc.

KAP Wrap

Look back at the Test Practice. Choose one problem you got wrong and explain the mistake you made.

What problem did you choose? _____

On the lines below, explain the mistake you made.

_____

_____

_____

Solve the problem correctly in the space below.

The 4-Step Method for Problem Solving

© 2006 Kaplan, Inc.

Thinking KAP

Here are the instructions for playing a fantasy adventure video game:

> The Elven queen, Maplewood, has been captured by an evil king. As the elf hero, Oakthorn, your first task is to find an enchanted ring that will allow you to enter the castle in which she is being held. To find the ring, you will need to journey north, battling 10 foes, including the invisible Grinkles, the venomous Croplunkers, and the giant Kelkies. Then you need to cross the River of Dreams, where you will need to hold your breath for the entire journey or fall asleep forever. Above all, never forget this: *the enchanted ring can only be seen by one whose eyes are closed.*
>
> **Good luck!**

What is important to know about playing this video game? How do you know?

Being a Math Detective

In the Thinking KAP activity, the directions held many clues about how to play the video game. Math problems contain clues, too. Good problem solvers look for clues that will help them understand a math problem.

In this lesson, you will work on thinking through math problems before you begin solving them. If you follow the 4-Step Method for Problem Solving, you'll always know the right questions you need to ask about each problem. Use the method on every problem on the ACT Mathematics Test.

The 4-Step Method for Problem Solving

Every time you see the finger symbols beside each step of the 4-Step Method for Problem Solving, you will be reminded of the four steps.

STEP 1: Understand the problem.

STEP 2: Analyze important information.

STEP 3: Plan and solve.

STEP 4: Check your work.

© 2006 Kaplan, Inc.

Getting an Understanding

Your first job is to understand the problem. You need to determine what kind of math the problem is about and what you are being asked to find.

👉 STEP 1: Understand the problem.

- **Scan the problem.**

- **Restate the question in your own words.**

Scan the problem

The first thing you should do when you see a math problem is to scan it. When you scan, you get a general sense of what the problem is about. What jumps out at you? What are your first impressions? What seems familiar about this problem?

Restate the question in your own words

To make sure you understand the question you are being asked to answer, put the question in your own words.

> **keep in mind**
>
> In Step 1, just get a general sense of what the problem is about.

Try It Out! ✏️

1. Each student in a class is either a sophomore, junior, or senior. There are 15 students in the class. If there are twice as many sophomores as juniors, and there are 3 seniors, how many juniors are in the class?

What is the problem basically about?

find how many Juniors and Sophomores are in the class
finding something

Restate the question in your own words.

How many Juniors in the Class?

© 2006 Kaplan, Inc.

Searching for Clues

Once you understand what you are being asked to find, you need to organize the information you have. Read the problem again carefully, and identify the clues.

STEP 2: *Analyze important information.*

- *Underline the clues.*

- *Write the important information in a helpful way.*

keep in mind

Clues tell you about the math in the problem, not the story.

Try It Out!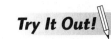

Underline the clues that tell you about the important information in the problem. Then fill in the table below.

1. Each student in a class is either a sophomore, junior, or senior. There are <u>15 students</u> in the class. If there are twice as many sophomores as juniors, and <u>there are 3 seniors</u>, how many juniors are in the class?

Clue	What does the clue tell you?
15 Students	How many Students in the class?
3 seniors	How many seniors in that class?

Rewriting the Important Information

For a long problem, it can be helpful to write the important information in a helpful way.

Draw It

One way to rewrite the important information is by drawing. Draw a simple figure to help you understand the information you are given.

Chart It

Another way to organize information is to make a chart. Charts can help you organize information in problems with:

- parts and totals
- categories of information
- changes over time

When creating a chart, include a column for the totals wherever possible.

keep in mind

Drawing and charting are just two ways to write the information in a helpful way. In this program, you will learn other ways, too.

Try It Out!

For the problem on the previous page, draw a picture of the important information. Then make a chart.

15 Students
3 seniors
_____
12 Students
8 sophomeres
4 Juniors

Which method of writing the important information did you find more helpful? Why?

_____

_____

The 4-Step Method for Problem Solving

STEP 1: Understand the problem.

■ *Scan the problem.*

■ *Restate the question in your own words.*

STEP 2: Analyze important information.

■ *Underline the clues.*

■ *Write the important information in a helpful way.*

STEP 3: Plan and solve.

■ *What do you know?*

■ *What do you need?*

■ *What can you do?*

STEP 4: Check your work.

■ *Did you answer the right question?*

■ *Is your answer reasonable?*

■ *Can you solve the problem a different way?*

© 2006 Kaplan, Inc.

Apply Steps 1 and 2 of the 4-Step Method for Problem Solving.

1. The circumferences of two circles have lengths in the ratio of 4:9. If the diameter of the smaller circle is 9 inches long, how many inches long is the diameter of the larger circle?

STEP 1: Understand the problem. *about ratios*

What is the problem basically about? *finding the diameter lrge circle*

Restate the question in your own words. *What is the diameter of the large circle?*

STEP 2: Analyze important information.

Underline the clues in the problem.

- What does the clue *ratio of 4:9* tell you? *9 inches long ...es of two circles*

- What does the clue *diameter of the smaller circle* tell you? *how long the diame of the small circle is*

Write the important information in a helpful way.

© 2006 Kaplan, Inc.

2. The table below shows the high temperature, in degrees Fahrenheit, on a certain day for seven cities. If the median high temperature of these cities is the same as Covington's high temperature on that day, which of the following could NOT be Covington's high temperature on that day?

City	High Temperature
Allentown	90°
Baylor	82°
Covington	?
Dayton	78°
Eaton	100°
Fernwood	80°
Glenview	88°

F. 82

G. 85

H. 86

J. 88

K. 89

 STEP 1: Understand the problem.

What is the problem basically about? _____

Restate the question in your own words. _____

STEP 2: Analyze important information.

Underline the clues in the problem.

- What does the clue *median* tell you? _____

- What do the clues in the table tell you? _____

Write the important information in a helpful way. _____

_____

© 2006 Kaplan, Inc.

Shared Practice

Solve the following problems. Don't forget to use Steps 1 and 2 of the 4-Step Method for Problem Solving to help you understand the problem and analyze the important information before you solve.

1. Two large sodas contain the same amount as 3 medium sodas. Two medium sodas contain the same amount as 3 small sodas. How many small sodas contain the same amount as 8 large sodas?

   A. 12

   B. 16

   C. 18

   D. 24

   E. 36

   hint *Make a chart to help you analyze the clues.*

2. Johanna purchases $250 worth of stock. The price of the stock goes up 20% in the next week. The price then falls by 20% on the following week. If Johanna then decides to sell the stock, what will be her net earning or loss from the transactions?

   F. She loses $10.

   G. She loses $4.

   H. She breaks even.

   J. She earns $5.

   K. She earns $10.

   hint *Extend the chart to help you answer the question.*

3.   A telephone poll was conducted to gauge the popularity of a potential senatorial candidate. Of the 800 people surveyed, 440 had a positive impression, 240 had a negative impression, and the rest had no opinion. What percent of those surveyed had no opinion of the potential candidate?

A.   12%

B.   15%

C.   30%

D.   85%

E.   120%

hint   *Which answer choices can you eliminate?*

4.   The table below, describing College X's freshman class population, is partially filled in. Based on the information in the table, how many female freshman students reside off-campus?

	On-campus residents	Off-campus residents	Total
Male	305		515
Female			
Total	517		835

F.   108

G.   110

H.   208

J.   210

K.   212

hint   *To write the information in a helpful way, mark the table to show the information you need.*

© 2006 Kaplan, Inc.

5. To keep up with rising commercial property taxes, a parking lot owner needs to raise the monthly parking rate of $130 by 25%. What will be the new monthly parking rate?

    A. $130.25
    B. $137.50
    C. $155.00
    D. $162.50
    E. $195.00

    hint Which answer choices are clearly too small or too large?

6. The average of a set of six integers is 14. If a seventh integer is added to the set, the average rises to 16. What is the seventh integer?

    F. 18
    G. 20
    H. 24
    J. 28
    K. 30

    hint How much more than 16 does the seventh integer have to be to raise each of the other 6 scores by 2 points?

© 2006 Kaplan, Inc.

7. Four friends were about to pay $20.00 each to enter an amusement park when they discovered that groups of 5 can pay a special group entrance fee for $85.00. How much would each of the friends save if they could convince a fifth person to join them and the 5 people divided equally the price of the special group entrance fee?

A. $ 1.50
B. $ 2.50
C. $ 3.00
D. $10.00
E. $12.00

hint ▷ *What's the significance of the word "each" here?*

8. John wants to make fruit salad. He has a recipe that serves 6 people and uses 4 oranges, 5 pears, 10 apples, and 2 dozen strawberries. If he wants to serve 18 people, how many pears will John need?

F.   11
G.   15
H.   18
J.   24
K.   30

hint ▷ *Read the question carefully and don't bother keeping track of information that you won't need.*

© 2006 Kaplan, Inc.

KAP Wrap

Choose any problem from the Shared Practice. Explain how you wrote the important information in a helpful way. Did you draw a picture? Make a chart? Make a list? Label a diagram? Why did you choose this strategy?

_____

_____

In the space below, try to organize the information a different way.

© 2006 Kaplan, Inc.

lesson B *Taking Action*

Thinking KAP

You have just reached level 3 of a fantasy adventure video game. Journeying on your way to find an enchanted ring, you reach a large lake. You must get to the other side of the lake in order to continue your quest. Unfortunately, your character, the elf hero Oakthorn, does not know how to swim!

What can you do? List as many options as you can.

© 2006 Kaplan, Inc.

Strategy Instruction

Making a Plan

There are many ways Oakthorn can get across the lake. There are many ways to solve a math problem, too. Good problem solvers know a lot of strategies for solving problems. They also know how to choose the right strategy for the job.

Once you understand a math problem and have analyzed the important information, you need to make a plan you can use to solve it.

For a multi-step problem, you may not have all the information you need right away. You need to solve part of the problem to get the information you need.

STEP 3: Plan and solve.

- **What do you know?**
- **What do you need?**
- **What can you do?**

Try It Out!

Use the three planning questions to solve the problem below.

1. In the figure below, $\overline{AB} = ?$

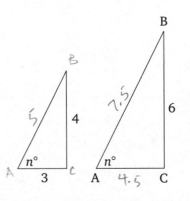

- **What do you know?** What information is in the problem? What background information do you bring to the problem?

- **What do you need?** What information will help you answer the question or move you toward the solution?

- **What can you do?** How can you use what you know to find what you need?

© 2006 Kaplan, Inc.

Checking Your Work

Once you have solved the problem, you are not finished! A math problem is not really solved if it's not checked. There are three questions you can use to check your work.

 **STEP 4: Check your work.**

- *Did you answer the right question?*

- *Is your answer reasonable?*

- *Can you solve the problem another way?*

Did you answer the right question?

One common reason test takers get problems wrong on Test Day is that they do not answer the question asked in the problem. After you solve a problem, go back and recall the question you restated in your own words in Step 1. Is this the question you have answered?

keep in mind

To answer the first check question, you will need to go back to the work you did in Step 1.

Try It Out!

1. In the figure below, $\overline{AB} = $?

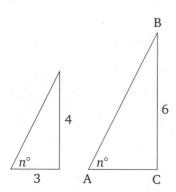

    (**A.** 4.5)
    B. 6
    C. 7.5
    D. 10
    E. 20

Reread the question in the problem above. What question has this student answered instead?

Is your answer reasonable?

Go back to the information you organized in Step 2. Does your answer make sense when you look at the drawing or chart? Is it reasonable when you compare it to what you already know?

Try It Out!

1.   In the figure below, $\overline{AB}$ = ?

keep in mind

To answer the second check question, you will need to go back to the work you did in Step 2.

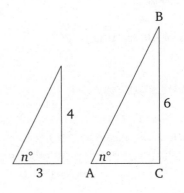

A.   4.5
B.   6
C.   7.5
D.   10
E.   20

Compare this answer to the lengths given in the problem. Is the answer reasonable?

© 2006 Kaplan, Inc.

Can you solve the problem another way?

In Step 3, you planned a problem-solving strategy and carried out your plan. Solving the problem another way can help you convince yourself that your answer is correct. If you solve the problem another way and don't get the same answer, you might want to check your computation more carefully.

 **Try It Out!**

1. In the figure below, $\overline{AB}$ =?

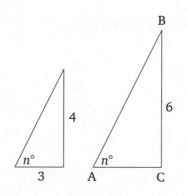

keep in mind

Most questions on the ACT Mathematics Test can be solved more than one way.

A.   4.5

B.   6

C.   7.5

D.   10

E.   20

My work:

The small triangle is a 3-4-5 triangle.

$$\frac{4}{5} = \frac{6}{x}$$

$$4x = 30$$

$$x = 7.5$$

This student used a proportion to solve the problem. Check this student's work by substituting the measurements of the larger triangle into the Pythagorean formula. How does this help you determine whether the answer is right or wrong?

© 2006 Kaplan, Inc.

The 4-Step Method for Problem Solving

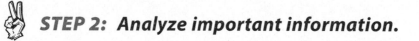 **STEP 1:** *Understand the problem.*

STEP 2: *Analyze important information.*

STEP 3: *Plan and solve.*

STEP 4: *Check your work.*

© 2006 Kaplan, Inc.

1. If $2a = 3b$, $4b = 5c$, and $2a - 6 = 24$, then $c = ?$

   A. 6
   B. 8
   C. 10
   D. 12
   E. 16

$c = ?$

$2a = 3b$

$4b = 5c$

$2(15) = 3b$

$30 = 3b$
$\overline{3} \quad \overline{3}$

$10 = b$

$4(10) = 5c$

$\dfrac{40}{5} = \dfrac{5c}{5}$

$8 = c$

$2a - 6 = 24$
$+6 \quad +6$

$\dfrac{2a}{2} = \dfrac{30}{2}$

$a = 15$

$3b - 6 = 24$
$+6 \quad 6$

$\dfrac{3b}{3} = \dfrac{30}{3}$

$b = 10$

© 2006 Kaplan, Inc.

2. In the standard (x, y) coordinate plane, what is the area of a triangle with vertices at $(1, 2)$, $(^-6, 2)$ and $(1, ^-4)$?

   F.   12

   G.   15

   H.   21

   J.   24

   K.   28

B

© 2006 Kaplan, Inc.

Shared Practice

Use the 4-Step Method for Problem Solving to solve the following problems.

1. Jimmy's school is 1.2 miles from his house, on the same street. If he walks down the street directly from the school to his house, he passes a candy store and a grocery store. The candy store is 0.35 miles from the school, and the grocery store is 0.67 miles from the candy store. How many miles is it from the grocery store to Jimmy's house?

A. 0.18

B. 0.53

C. 0.83

D. 0.85

E. 1.92

hint ▶ *Make a drawing to help organize the information.*

2. A necklace is strung with beads of different colors in the following pattern: green, green, red, yellow, yellow, blue, white, white, red. If the pattern keeps repeating, what will be the color of the 57th bead on the string?

F. blue

G. green

H. red

J. yellow

K. white

hint ▶ *Determine what you know and what you need.*

© 2006 Kaplan, Inc.

3. Jude has 12 shirts, 3 of which are black. If he selects a shirt at random, what is the probability that it will NOT be black?

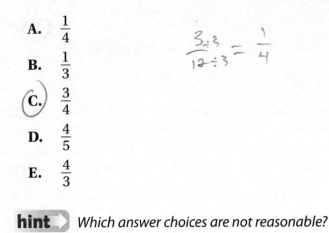

A. $\frac{1}{4}$

B. $\frac{1}{3}$

C. $\frac{3}{4}$

D. $\frac{4}{5}$

E. $\frac{4}{3}$

hint ▷ *Which answer choices are not reasonable?*

4. When a square label with a side length of 8 inches is wrapped around a right cylindrical can, the sides meet with no overlap. What is the area, in square inches, of the base of the can?

F. 16π

G. 8π

H. 4π

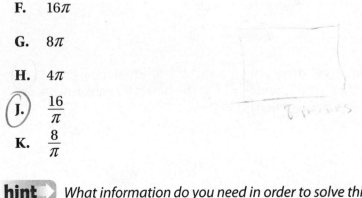

J. $\frac{16}{\pi}$

K. $\frac{8}{\pi}$

hint ▷ *What information do you need in order to solve this problem? What do you know from the information given?*

B

© 2006 Kaplan, Inc.

5. If $3x + 4 = {}^-5$, what is the value of $x^2 - 4x$?

    A. $^-3$
    B. 3
    C. 9
    D. 21
    E. 45

$3x + 4 = {}^-5$
$\quad -4 \quad -4$
_____
$\dfrac{3x}{3} = \dfrac{-9}{3}$

$x = {}^-3$

$(^-3)^2 - 4(^-3)$

$9 + 12 = 21$

    hint Begin with what you know.

6. What is the length in units of one side of a square with perimeter $16 - 24h$ units?

    F. $16 - 24h$
    G. $16 - 6h$
    H. $8h$
    J. $4 - 24h$
    K. $4 - 6h$

$\dfrac{16 - 24h}{4}$

$4 - 6h$

    hint What do you know about the relationship between a side of a square and the perimeter of the square?

© 2006 Kaplan, Inc.

7. A certain mixture of trail mix contains peanuts, raisins, and sesame sticks in a ratio of 2:4:5 by weight. How many ounces of the mixture can be made with 10 ounces of peanuts?

A. 21

B. 22

C. 27.5

D. 40

E. 55

P R S
2 4 5

10 20 25

hint ▷ *You can use a chart to organize the information.*
Check your work by asking yourself if your answer
seems reasonable.

8. In the standard (x, y) coordinate plane, how many times does the graph of $y = (x + 1)(x + 2)(x - 3)(x + 4)(x + 5)$ intersect the x-axis?

F. 15

G. 9

H. 5

J. 4

K. 1

hint ▷ *What is special about points that intersect the x-axis?*

ACT ADVANTAGE
MATHEMATICS

© 2006 Kaplan, Inc.

KAP Wrap

Imagine you have a classmate who has been out of school for a few days and doesn't know about the 4-Step Method for Problem Solving. Write your classmate a letter explaining the method you have learned for solving math problems on the ACT Mathematics Test.

© 2006 Kaplan, Inc.

Putting It All Together

 ReKAP

Review the strategies from Lessons A and B. Then complete the following:

1. The four steps in the 4-Step Method for Problem Solving are:

_____

_____

_____

_____

2. Three questions I can ask myself to check my work are:

_____

_____

_____

© 2006 Kaplan, Inc.

Eliminating Unreasonable Choices

You will learn many strategies for understanding problems, organizing information, planning and solving problems, and checking your work. One useful strategy that you can use for many types of problems is Eliminating.

Eliminating allows you to make an educated guess. Sometimes it even helps you narrow the choices down to one.

Eliminating
• Cross out answer choices that are clearly wrong.
• Guess from among the remaining answer choices.

Try It Out!

1. A geologist weighs five rocks. Their weights are 12.38 kilograms, 12.46 kilograms, 14.07 kilograms, 14.11 kilograms, and 14.68 kilograms. What is the average weight of the five rocks, in kilograms?

    A. 11.21

    B. 12.36

    C. 13.54

    D. 14.72

    E. 15.91

Which answer choices can you eliminate without calculating the answer?

© 2006 Kaplan, Inc.

Visually Estimating Answers

If the question provides a figure, or if you can draw one from the information provided, you can often estimate the correct choice and eliminate unreasonable answer choices.

Try It Out!

2. In the figure below, $\overline{BD}$ is the altitude of an equilateral triangle $\triangle ABC$. If $\overline{BD}$ is $12\sqrt{3}$ units long, how many units long is $\overline{AC}$?

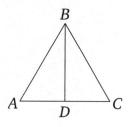

keep in mind

You can often visually estimate lengths. Is $\overline{AC}$ longer or shorter than $\overline{BD}$? Is it by a lot or by a little?

F. $6\sqrt{3}$

G. 12

H. $12\sqrt{3}$

J. 24

K. $24\sqrt{3}$

© 2006 Kaplan, Inc.

Test Practice Unit 2

When your teacher tells you, carefully tear out this page. Then begin working.

1. Ⓐ Ⓑ Ⓒ Ⓓ Ⓔ   11. Ⓐ Ⓑ Ⓒ Ⓓ Ⓔ

2. Ⓕ Ⓖ Ⓗ Ⓙ Ⓚ   12. Ⓕ Ⓖ Ⓗ Ⓙ Ⓚ

3. Ⓐ Ⓑ Ⓒ Ⓓ Ⓔ   13. Ⓐ Ⓑ Ⓒ Ⓓ Ⓔ

4. Ⓕ Ⓖ Ⓗ Ⓙ Ⓚ   14. Ⓕ Ⓖ Ⓗ Ⓙ Ⓚ

5. Ⓐ Ⓑ Ⓒ Ⓓ Ⓔ   15. Ⓐ Ⓑ Ⓒ Ⓓ Ⓔ

6. Ⓕ Ⓖ Ⓗ Ⓙ Ⓚ

7. Ⓐ Ⓑ Ⓒ Ⓓ Ⓔ

8. Ⓕ Ⓖ Ⓗ Ⓙ Ⓚ

9. Ⓐ Ⓑ Ⓒ Ⓓ Ⓔ

10. Ⓕ Ⓖ Ⓗ Ⓙ Ⓚ

C

© 2006 Kaplan, Inc.

MATHEMATICS TEST
15 Minutes—15 Questions

DIRECTIONS: Solve each problem, choose the correct answer, and then fill in the corresponding oval on your answer document.

Do not linger over problems that take too much time. Solve as many as you can; then return to the others in the time you have left for this test.

You are permitted to use a calculator on this test. You may use your calculator for any problems you choose, but some of the problems may best be done without using a calculator.

Note: Unless otherwise stated, all of the following should be assumed.

1. Illustrations are NOT necessarily drawn to scale.
2. Geometric figures lie in a plane.
3. The word *line* indicates a straight line.
4. The word *average* indicates arithmetic mean.

---

1. A rod is a unit of length, equivalent to 5.5 yards. If a field is 127 yards long, then how many rods long is the field, to the nearest tenth?

   A.   231.9
   B.   69.9
   C.   43.3
   D.   23.1
   E.   4.3

2. A company conducted a taste test of its new soft drink. Of the 1,250 participants, 800 liked the soft drink. Of the remaining participants, 40% disliked the soft drink and the rest were undecided. How many participants in the taste test were undecided about the new soft drink?

   F.   180
   G.   270
   H.   320
   J.   500
   K.   750

DO YOUR FIGURING HERE.

GO ON TO THE NEXT PAGE.

© 2006 Kaplan, Inc.

3. In a company with 350 employees, 120 of the employees are women. If 20% of the employees work in the accounting department, and 7 women work in the accounting department, how many employees are men who do NOT work in the accounting department?

A. 13
B. 70
C. 113
D. 167
E. 230

4. If the ratio of males to females in a group of students is 3:5, which of the following could be the total number of students in the group?

F. 148
G. 150
H. 152
J. 154
K. 156

5. Joseph can install a driveway in 9 hours. Mitch can install a driveway in 18 hours. If they work together at their individual rates, how many hours would it take them to install a driveway?

A. 4.5
B. 6
C. 9
D. 13.5
E. 27

GO ON TO THE NEXT PAGE.

© 2006 Kaplan, Inc.

6. Train A travels 50 miles per hour for 3 hours; Train B travels 70 miles per hour for $2\frac{1}{2}$ hours. What is the difference between the number of miles traveled by Train A and the number of miles traveled by Train B ?

F.   0

G.   25

H.   150

J.   175

K.   325

DO YOUR FIGURING HERE.

7. If the perimeter of parallelogram *RSTU* shown below is 42 units, how many units long is $\overline{UT}$?

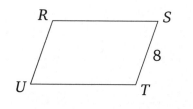

A.   13

B.   15.25

C.   21

D.   26

E.   34

8. Which of the following is a value of *b* for which $(b - 3)(b + 4) = 0$?

F.   ⁻3

G.   0

H.   3

J.   4

K.   6

GO ON TO THE NEXT PAGE.

9. What is the length, in inches, of the hypotenuse of a right triangle with legs measuring 8 inches and 15 inches?

   A. $\sqrt{23}$
   B. $\sqrt{161}$
   C. 7
   D. 17
   E. 23

10. It takes 5 eggs to make 20 muffins that each weigh 3 ounces. Using the same recipe, how many eggs would it take to make 14 muffins that each weigh 6 ounces?

    F. 3
    G. 5
    H. 7
    J. 9
    K. 11

11. At a certain store, item A costs $4 more than item B, and item B costs twice as much as item C. If m is the cost of item C in dollars, what is the cost of item A in dollars?

    A. $m + 4$
    B. $m + 8$
    C. $2m + 4$
    D. $2m + 8$
    E. $4m$

GO ON TO THE NEXT PAGE.

ACT ADVANTAGE
MATHEMATICS

© 2006 Kaplan, Inc.

12. The relative atomic mass of an element is the ratio of the mass of the element to the mass of an equal amount of carbon. If 1 cubic centimeter of carbon has a mass of 12 grams, what is the relative atomic mass of an element that has a mass of 30 grams per cubic centimeter?

F. 1

G. 1.2

H. 2.5

J. 3

K. 30

13. When $\frac{4}{11}$ is written as a decimal, what is the 100th digit after the decimal point?

A. 3

B. 4

C. 5

D. 6

E. 7

14. Pat deposited 15% of last week's take-home pay into a savings account. If she deposited $37.50, what was last week's take-home pay?

F. $ 25.00

G. $ 56.25

H. $112.50

J. $225.00

K. $250.00

GO ON TO THE NEXT PAGE.

© 2006 Kaplan, Inc.

15. $\left(\dfrac{1}{5} + \dfrac{1}{3}\right) + \dfrac{1}{2} = ?$

A. $\dfrac{1}{8}$

B. $\dfrac{1}{4}$

C. $\dfrac{4}{15}$

D. $\dfrac{31}{30}$

E. $\dfrac{16}{15}$

END OF TEST.

STOP! DO NOT TURN THE PAGE UNTIL TOLD TO DO SO.

© 2006 Kaplan, Inc.

KAP Wrap

You have now learned the entire 4-Step Method for Problem Solving.
For each step, list the most helpful strategy you learned in Unit 2. What
mistakes do you think the strategy can help you avoid? How can it help you
solve problems more easily?

 STEP 1: Understand the problem.

 STEP 2: Analyze important information.

 STEP 3: Plan and solve.

 STEP 4: Check your work.

© 2006 Kaplan, Inc.

Unit 3
Pre-Algebra

© 2006 Kaplan, Inc.

lesson A *Number Properties and Central Tendency*

Thinking KAP

On Level 5 of a fantasy adventure video game, you notice that your power level has gotten very low. You need to get 8 more power points soon, or the game will be over. If you find and pick up 2 green flasks and 1 red jewel, you will get exactly 8 power points.

How many power points could the green flask be worth? How many power points could the red jewel be worth? How did you find your answer?

Picking Numbers

To complete the Thinking KAP activity, you may have picked a value for the green flask and used that value to determine how many power points the red jewel was worth. This is similar to a strategy called Picking Numbers. You can use this strategy to answer questions about number properties.

Picking Numbers lets you solve abstract problems in concrete ways.

Picking Numbers
• Pick a number for an unknown number or variable. Do not pick 1 or 0.   • Use the number to compute an answer to the problem.   • Substitute the number you picked for the unknown value in each answer choice. Eliminate answer choices that do not yield the correct answer.

Try It Out!

1. Which of the following calculations will yield an odd integer for any integer n?

    A. $n + 1$
    B. n^2
    C. $n^2 + 1$
    D. $2n^2 + 1$
    E. $2n^2 + 2$

Pick 2, an even number, for n. Because you know the correct choice will yield an odd integer, substitute and solve to see which answer choices are odd when n equals 2.

Eliminate choices that are not odd integers. Then pick an odd integer and substitute to eliminate remaining choices that are not odd integers.

© 2006 Kaplan, Inc.

Exponents

Some math problems require you to use laws of exponents, like the ones shown in the table below.

Rule	Examples
$x^0 = 1$	$v^0 = 1$ $3^0 = 1$
$x^m \cdot x^n = x^{m+n}$	$z^2 \cdot z^3 = z^5$ $2^2 \times 2^3 = 2^5 = 32$
$x^1 = x$	$5^1 = 5$
$\dfrac{x^m}{x^n} = x^{m-n}$	$\dfrac{b^8}{b^2} = b^6$ $\dfrac{5^4}{5} = 5^3 = 125$
$x^{-n} = \dfrac{1}{x^n}$	$g^{-4} = \dfrac{1}{g^4}$ $7^{-2} = \dfrac{1}{7^2} = \dfrac{1}{49}$
$(x^m)^n = x^{mn}$	$(p^3)^3 = p^9$ $(3^2)^3 = 3^6 = 729$
$(xy)^n = x^n y^n$	$(jk)^8 = j^8 k^8$ $(4r)^2 = 4^2 \times r^2 = 16r^2$
$\left(\dfrac{x}{y}\right)^n = \dfrac{x^n}{y^n}$	$\left(\dfrac{a}{b}\right)^6 = \dfrac{a^6}{b^6}$ $\left(\dfrac{2}{3}\right)^2 = \dfrac{2^2}{3^2} = \dfrac{4}{9}$

keep in mind

The only way to *add* or *subtract* expressions with exponents is to factor out like terms and add what is left. For instance, $x^2 + x^2 = 1x^2 + 1x^2 = (1 + 1)x^2 = 2x^2$.

Try It Out!

2. For all $x \neq 0$, $\dfrac{x^2 + x^2 + x^2}{x^2} = ?$

   F. 3

   G. $3x$

   H. x^2

   J. x^3

   K. x^4

Which law of exponents can help you solve this problem?

_____

Check your work with Picking Numbers.

© 2006 Kaplan, Inc.

All Things Equal

Some problems ask you to simplify expressions. You can solve these problems by using pencil and paper, but you should also consider using All Things Equal.

All Things Equal
• To compare expressions, use your calculator to convert each expression into a numerical value.

keep in mind

When adding radicals on a calculator, put each radical in parentheses. For example, to add $\sqrt{50} + \sqrt{8}$, enter it as $(\sqrt{50}) + (\sqrt{8})$. Doing this tells the calculator to take the square root of each number, not of the sum.

Try It Out!

3. Which of the following is equivalent to $\sqrt{50} + \sqrt{8} + \sqrt{18}$?

    A. 12
    B. $10\sqrt{2}$
    C. $4\sqrt{19}$
    D. 20
    E. 38

Use All Things Equal.

- Find the sum of $\sqrt{50} + \sqrt{8} + \sqrt{18}$ on your calculator. _____

- Is the sum a decimal or a whole number? Which answer choices can you eliminate? _____

- Use your calculator to find the numerical values of the remaining choices. Which choice has the same value as the sum? _____

A

© 2006 Kaplan, Inc.

Measures of Central Tendency

You can use memory devices to help you recall the differences among these three measures. The examples in the table below apply to the following set of data: 30, 40, 10, 60, 10.

Memory Device	Definition	Example
Mean means *average*.	The mean is the sum of the terms divided by the number of terms.	$\text{Mean} = \dfrac{30 + 40 + 10 + 60 + 10}{5} = 30$
Median sounds like *middle*.	The median is the middle number when the data is ordered from least to greatest.	From least to greatest: 10, 10, <u>30</u>, 40, 60. So, the median is 30.
Mode sounds like *most*.	The mode is the number that appears most often in a data set. (see above rows)	The mode is 10 because it appears twice in the data set, and the other terms appear only once.

keep in mind

If there is an even number of terms in a data set, average the two middle numbers to find the median.

Try It Out!

The scores that Reana received on her last four physics quizzes are 74, 82, 72, and 82. Find the mean, median, and mode of her quiz scores.

- mean: _____

- median: _____

- mode: _____

If Reana earns a 90 on her next physics quiz, how will her mean score change?

_____

_____

© 2006 Kaplan, Inc.

A

The 4-Step Method for Problem Solving

 STEP 1: *Understand the problem.*

STEP 2: *Analyze important information.*

STEP 3: *Plan and solve.*

STEP 4: *Check your work.*

Picking Numbers

- Pick a number for an unknown number or variable. Do not pick 1 or 0.
- Use the number to compute an answer to the problem.
- Substitute the number you picked for the unknown value in each answer choice. Eliminate answer choices that do not yield the correct answer.

All Things Equal

- To compare expressions, use your calculator to convert each expression into a numerical value.

© 2006 Kaplan, Inc.

1. If $5x^4y^5 > 0$, then which of the following CANNOT be true?

   A. $x < 0$
   B. $x > 0$
   C. $x = y$
   D. $y < 0$
   E. $y > 0$

2. Martin's average score on 4 tests is 89. What score would Martin have to earn on the fifth test to raise his average score to 90?

F. 90

G. 91

H. 92

J. 93

K. 94

© 2006 Kaplan, Inc.

Shared Practice

Use the 4-Step Method for Problem Solving and the strategies you have learned in this lesson to solve the problems in this section.

1. Which of the following is equivalent to $\dfrac{8\frac{1}{3}}{6\frac{2}{3}}$?

    A. $1\frac{1}{4}$

    B. $1\frac{2}{3}$

    C. $2\frac{1}{2}$

    D. $2\frac{2}{3}$

    E. $55\frac{4}{9}$

    hint ▷ *Use All Things Equal. Convert the fraction in the question and the fractions in the answer choices to decimals and compare.*

2. Janice and her friends are playing a game with 60 cards. Each player has an equal number of cards, and there are no remaining cards. If a player were to leave, the cards could be redealt such that each player would still get an equal number of cards with no cards remaining. What is the greatest possible number of players who can participate in the game?

    F. 5

    G. 6

    H. 12

    J. 15

    K. 20

    hint ▷ *Pay attention to the clue "greatest" in this question. It is not enough to find a possible number of people who can play the game—you must find the largest possible number.*

© 2006 Kaplan, Inc.

3. If *m* is an integer, then $(m - 1)^2 + 3$ must be:

    A. an odd integer.
    B. an even integer.
    C. an integer divisible by 3.
    D. a positive integer.
    E. a negative integer.

     Pick one or more numbers for m and substitute it into the expression to see which answer choice is always true.

4. Rachel's average score after 6 tests is 83. If Rachel scores a 97 on the seventh test, which of the following statements must be true?

    F. Her mean test score will increase by 2 points.
    G. Her mean test score will increase by 14 points.
    H. Her median test score will decrease by 2 points.
    J. Her median test score will increase by 14 points.
    K. Her mode test score will stay the same.

    hint *Use memory devices to help you recall the differences between the mean, median, and mode of a set of data.*

A

© 2006 Kaplan, Inc.

5. If an integer is divisible by 6 and by 9, then the integer must also be divisible by which of the following?

        I. 12

        II. 18

        III. 36

    A. I only

    B. II only

    C. I and II only

    D. I, II, and III

    E. None

> **hint** ▷ *Pick several numbers that are divisible by both 6 and 9, such as 18 and 54. See if those numbers are evenly divisible by 12, 18, and 36.*

6. What is the value of $(^-2)^{-3} + (^-3)^{-2}$?

    F. $-\dfrac{17}{72}$

    G. $-\dfrac{1}{72}$

    H. 0

    J. $\dfrac{1}{72}$

    K. $\dfrac{17}{72}$

> **hint** ▷ *Use All Things Equal to convert the expression and each answer choice to the same form.*

© 2006 Kaplan, Inc.

7. What is the average of the expressions $2x + 5$, $5x - 6$, and $^-4x + 2$?

A. $x + \dfrac{1}{3}$

B. $x + 1$

C. $3x + \dfrac{1}{3}$

D. $9x + 3$

E. $3x + 3\dfrac{1}{3}$

hint ▸ Pick 3 for x and use it to find the average of the terms. Then substitute 3 into each answer choice to see which one is equivalent to that average.

8. If $x > 0$ and $y > 0$, $\dfrac{\sqrt{x}}{x} + \dfrac{\sqrt{y}}{y}$ is equivalent to which of the following?

F. $\dfrac{2}{\sqrt{xy}}$

G. $\dfrac{\sqrt{x} + \sqrt{y}}{xy}$

H. $\dfrac{x + y}{xy}$

J. $\dfrac{\sqrt{x} + \sqrt{y}}{\sqrt{x + y}}$

K. $\dfrac{\sqrt{x} + \sqrt{y}}{\sqrt{xy}}$

hint ▸ Pick numbers for x and y and use them to find the equivalent expression.

ACT ADVANTAGE
MATHEMATICS

© 2006 Kaplan, Inc.

KAP Wrap

Suppose one of your classmates was absent from school when Picking Numbers was taught. Choose a problem from the Shared Practice for which Picking Numbers would have been helpful. Then write a step-by-step explanation for your classmate, describing how Picking Numbers could be used to solve that problem.

A

© 2006 Kaplan, Inc.

lesson B Proportions and Probability

Thinking KAP

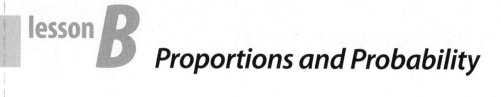

After playing video games for an hour, you and your friend agree to share a snack. You head to the refreshment area. Reaching into your pocket, you pull out a five-dollar bill and say, "I'll contribute 50% of this." Your friend pulls out a ten and a five and says, "Okay, I'll throw in 10% of what I have."

Snack Bar Menu

Soda			Popcorn		
	small	$0.75		small	$1.50
	medium	$1.25		medium	$1.75
	large	$1.50		large	$2.50
Chips		$0.75	Pretzel		$0.75
Gum		$0.50	Hot Dog		$1.50

Arriving at the snack bar, you glance up at the price list. What do you order?

© 2006 Kaplan, Inc.

Ratio Charts

In the Thinking KAP activity, you solved the problem by using a type of ratio called a percent. A ratio is a comparison of two quantities. For example, the ratio of boys to girls on a class trip might be 1 to 3. This ratio can be written using a colon, 1:3, or using a fraction, $\frac{1}{3}$. Ratios can be used to express part-to-part or part-to-whole relationships.

A **proportion**, such as, $\frac{1}{3} = \frac{2}{6}$, may be used to express equivalent ratios. A Ratio Chart can also show equivalent ratios.

A rate is another special type of ratio that compares two different units, such as miles and hours. For example, 60 miles per hour is a rate.

Ratio Charts
• Create a Ratio Chart and label the rows.
• Fill in as many cells as you can with the data available.
• Fill in missing information.

 *Try It Out!*

1. The ratio of green apples to red apples in a basket is 3:4. If there is a total of 21 apples, how many green apples are there?

	Ratio 1	Ratio 2
green apples	3	
red apples	4	
total apples		21

© 2006 Kaplan, Inc.

Percents and Ratio Charts

You can also use Ratio Charts to solve percent problems. A percent expresses a part-to-whole relationship—the ratio of a number to 100.

To solve a multi-step percent problem, such as a problem involving a percent increase or decrease, you may need to use a Ratio Chart to complete the first task, and then perform a computation to complete the second task.

Try It Out!

2. The regular price of a certain bicycle is $125.00. If that price is reduced by 20%, what is the new price?

    A. $100.00

    B. $105.00

    C. $112.50

    D. $120.00

    E. $122.50

keep in mind

Clue words that signal percent *increases* include "sales tax" and "commission." Clue words such as "discount," "reduced," or "percent off" usually signal percent *decreases*.

	Percent Ratio	**Original Price**
part		
whole		

Use the Percent Ratio column to express the percent. Use "100" for the whole. Then solve for the price reduction, and use that information to solve the problem.

Probability and Ratio Charts

A **probability** is the likelihood that an event will happen. It can be expressed as a part-to-whole ratio.

$$\text{probability} = \frac{\text{number of favorable outcomes}}{\text{total possible outcomes}}$$

For example, if the chance of winning a raffle is 1 in 100, the probability can be expressed as $\frac{1}{100}$. Probability ranges from 0 to 1. A probability of 0 means that it is impossible for an event to occur, and a probability of 1 means that it is certain that an event will occur.

Ratio Charts can be used to solve problems involving probabilities.

A probability is always a number from 0 to 1.

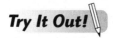

Try It Out!

3. Deshaun has a bag that contains only 6 blue marbles. He wants to add enough green marbles so that the probability of randomly selecting a blue marble from the bag is $\frac{1}{4}$. If he adds only green marbles, how many should he add?

	Probability Ratio	**Number of Marbles**
blue marbles		
total marbles		

Use the Ratio Chart to solve the problem below.

- Write the known information into the Ratio Chart.

- Use a number relationship to find the total number of marbles.

- Use this total to find the number of green marbles that should be added.

© 2006 Kaplan, Inc.

Combinations, Permutations, and Compound Probability

Some problems involve more than one event. For example, a problem may ask for the probability that two or more events will occur. Making an organized list can help you solve some of these problems.

For example, suppose you flip a fair coin and spin a spinner with three equal-sized sections, labeled *blue*, *red*, and *yellow*. You can make a list to show all the possible outcomes.

heads & blue	heads & red	heads & yellow
tails & blue	tails & red	tails & yellow

So, there are 6 possible outcomes. The probability of the coin landing heads-up and the spinner pointing to yellow is $\frac{1}{6}$ because the favorable outcome "heads – yellow" appears 1 time out of the 6 possible outcomes.

Another way to solve this type of problem is to multiply the probability of each independent event. The probability of the coin landing on heads is $\frac{1}{2}$. The probability of the spinner landing on yellow is $\frac{1}{3}$. So, the probability that the coin will land heads-up and the spinner will point to yellow is:

$$\frac{1}{2} \times \frac{1}{3} = \frac{1}{6}$$

keep in mind

Making lists and multiplying can also help you solve certain problems involving permutations (in which order matters) and combinations (in which order does not matter).

Try It Out!

Solve this problem in the space below.

4. Jaritza spins a spinner with 2 equal-sized sections, labeled orange and green. At the same time, she rolls a 6-sided dice numbered 1 through 6. What is the probability that the spinner will land on green and the die will land on a number greater than 4?

Did you make a list or use multiplication? Explain why you chose the strategy that you did.

_____

_____

B

The 4-Step Method for Problem Solving

 **STEP 1:** *Understand the problem.*

STEP 2: *Analyze important information.*

STEP 3: *Plan and solve.*

STEP 4: *Check your work.*

Ratio Charts
• Create a Ratio Chart and label the rows.
• Fill in as many cells as you can with the data available.
• Fill in missing information.

© 2006 Kaplan, Inc.

1. Jan types at an average rate of 12 pages per hour. At that rate, how long will it take Jan to type 100 pages?

   A. 8 hours and 3 minutes

   B. 8 hours and 15 minutes

   C. 8 hours and 20 minutes

   D. 8 hours and 30 minutes

   E. 8 hours and $33\frac{1}{3}$ minutes

© 2006 Kaplan, Inc.

2. How many different four-digit lock combinations can be made using each of the digits 4, 5, 6, and 7 once?

F. 4

G. 6

H. 12

J. 24

K. 48

© 2006 Kaplan, Inc.

Shared Practice

Use the 4-Step Method for Problem Solving and the strategies you have learned in this lesson to solve the problems in this section.

1. What value of *x* solves the following proportion?

    $$\frac{2}{9} = \frac{x}{15}$$

    A. $2\frac{2}{5}$

    B. 3

    C. $3\frac{1}{3}$

    D. $4\frac{1}{3}$

    E. $5\frac{1}{2}$

    hint ▷ *Use a Ratio Chart or cross-multiply to solve for the unknown value, x.*

2. In a certain string ensemble, the ratio of men to women is 5:3. If there is a total of 24 people, how many women are there?

    F. 12

    G. 11

    H. 10

    J. 9

    K. 8

    hint ▷ *Whenever you see "ratio," try using a Ratio Chart.*

B

3. A jar contains 4 grape gum balls, 5 cherry gum balls, and 11 orange gum balls. If one gum ball is chosen at random, what is the probability that it will be grape?

    A. $\frac{1}{3}$

    B. $\frac{1}{4}$

    C. $\frac{1}{5}$

    D. $\frac{1}{16}$

    E. $\frac{1}{55}$

    hint *You'll need to determine the total number of gum balls while solving this problem.*

4. A menu offers 4 choices for the first course, 5 choices for the second course, and 3 choices for dessert. How many different meals, consisting of a first course, a second course, and a dessert, can be chosen from this menu?

    F. 12

    G. 24

    H. 30

    J. 36

    K. 60

    hint *Multiply or make a list to find the total number of possibilities.*

© 2006 Kaplan, Inc.

B

5. If the probability that it will rain on Tuesday is 0.6, what is the probability that it will NOT rain on Tuesday?

    A.   0.0

    B.   0.4

    C.   0.6

    D.   1.0

    E.   1.6

> **hint**   *A probability is always a number from 0 to 1. Which choice can you eliminate as obviously wrong?*

6. Pat deposited 15% of last week's take-home pay into a savings account. If she deposited $37.50, what was last week's take-home pay?

    F. $ 25.00

    G. $ 56.25

    H. $112.50

    J. $250.00

    K. $375.00

> **hint**   *Use a Ratio Chart to find the total earnings when $37.50 is the part.*

© 2006 Kaplan, Inc.

7. If a fair coin is tossed 3 times, what is the probability that the coin will land with the heads side facing up exactly once?
    (Note: in a fair coin toss the two outcomes, heads and tails, are equally likely.)

    A. $\dfrac{1}{8}$

    B. $\dfrac{1}{3}$

    C. $\dfrac{3}{8}$

    D. $\dfrac{1}{2}$

    E. $\dfrac{2}{3}$

    hint *Make an organized list to help you determine the answer.*

8. From 1970 to 1980, the population of City Q increased by 20%. From 1980 through 1990, the population increased by 30%. What was the combined percent increase for the period 1970–1990?

    F. 25%

    G. 26%

    H. 36%

    J. 50%

    K. 56%

    hint *Use Picking Numbers. Pick 100 for the population in 1970, and find the new population after a 20% increase. Then increase that population by 30%. By what percentage did the original population of 100 increase?*

© 2006 Kaplan, Inc.

KAP Wrap

Charts and lists are two ways that you can organize the information in a problem to help you solve it. Choose a problem from the Shared Practice and explain how you used a chart or a list to help you solve that problem.

B

Tables, Graphs, and Charts

ReKAP

Review the strategies from Lessons A and B. Then fill in the blanks with what you have learned.

1. If you substitute a number you've picked into each choice, and you cannot eliminate all of the choices, you should _____.

2. If you divide two numbers that have the same base but different exponents, you should _____ one exponent from the other.

3. To find the probability of multiple outcomes, you should _____ the probability of each outcome.

4. What devices can you use to remember the terms used on ACT problems for measures of central tendency?

   _____

   _____

   _____

C

Working with Graphs

Solving problems about tables and graphs is easy if you know where to look for clues. You know that part of Step 1 of the 4-Step Method for Problem Solving is to underline clues. When you see a table or graph, you should pay careful attention to three types of clues.

Pay attention to the key in a graph. It will also give critical information that will allow you to read the graph.

Words, Numbers, Shape

- Look at the **words** in the title, headings, and key.
- Look at the **numbers** in the table, scale, and axes.
- Look at the **shape** of the data.

 *Try It Out!*

1. In the graph below, in which year were twice as many women's jackets sold as children's jackets?

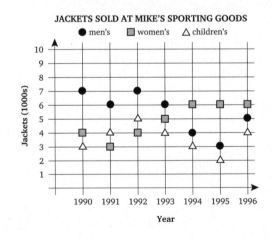

A. 1990

B. 1991

C. 1992

D. 1993

E. 1994

© 2006 Kaplan, Inc.

Working with Tables and Charts

Problems with tables and charts often require knowing the total amount represented by the table or chart. Be sure to locate that crucial piece of information. Also, look at the percentages of all the parts and make sure you pay attention to what you are being asked to solve.

2. How many cats were seen at the veterinary clinic last week?

**PETS SEEN AT THE
VETERINARY CLINIC LAST WEEK**

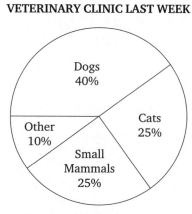

Total number of pets = 260

Problems with tables, charts, and graphs are like all other ACT Mathematics problems, except that the data is presented in a different format.

A.   25
B.   26
C.   65
D.   104
E.   130

Use Words, Numbers, Shape to identify the information you need. Then set up a Ratio Chart to solve the problem.

MATHEMATICS TEST

15 Minutes—15 Questions

DIRECTIONS: Solve each problem, choose the correct answer, and then fill in the corresponding oval on your answer document.

Do not linger over problems that take too much time. Solve as many as you can; then return to the others in the time you have left for this test.

You are permitted to use a calculator on this test. You may use your calculator for any problems you choose, but some of the problems may best be done without using a calculator.

Note: Unless otherwise stated, all of the following should be assumed.

1. Illustrations are NOT necessarily drawn to scale.
2. Geometric figures lie in a plane.
3. The word *line* indicates a straight line.
4. The word *average* indicates arithmetic mean.

---

1. What is the value of $(5\sqrt{3})^2$?

   A. 15
   B. $10\sqrt{3}$
   C. $25\sqrt{3}$
   D. 30
   E. 75

   DO YOUR FIGURING HERE.

   $25\sqrt{9}$ $25 \cdot 3 = 75$

2. Jamal has a suitcase that contains only 10 white socks. He wants to add enough black socks so that the probability of randomly selecting a white sock is $\frac{1}{5}$. How many black socks should Jamal add to the suitcase?

   F. 30
   G. 35
   H. 40
   J. 45
   K. 50

GO ON TO THE NEXT PAGE.

C

© 2006 Kaplan, Inc.

3. If $ab \geq 0$ and $3b \geq 1$, then which of the following must be true of $\frac{a}{b}$?

DO YOUR FIGURING HERE.

- **A.** $\frac{a}{b}$ must be positive.
- **B.** $\frac{a}{b}$ must be negative.
- **C.** $\frac{a}{b}$ must be greater than 1.
- **D.** $\frac{a}{b}$ must be less than 1.
- **E.** $\frac{a}{b}$ must be equal to 1.

4. Which of the following calculations will yield an even integer for any integer n?

- **F.** $2n + 1$ — 5
- **G.** $n^2 - 1$ — 2
- **H.** $n^2 + 1$ — 5
- **J.** $2n^2 + 1$ — 9
- **K.** $2n^2 + 2$ — 10

5. Mandy scored 150, 195, and 160 in three bowling games. What should she score on her next bowling game if she wants to have an average score of exactly 175 for the four games?

- **A.** 205
- **B.** 195
- **C.** 185
- **D.** 175
- **E.** 165

150   505X
195
160
X

6. After eating 25% of his jelly beans, Brett has 72 left. How many jelly beans did Brett have originally?

- **F.** 90
- **G.** 96
- **H.** 97
- **J.** 180
- **K.** 288

GO ON TO THE NEXT PAGE.

© 2006 Kaplan, Inc.

7. Alicia is playing a game in which she draws marbles from a box. There are 50 marbles, numbered 1 to 50. Alicia draws one marble from the box and sets it aside, then draws a second marble. If both marbles have the same units digit, then Alicia wins. If the first marble she draws is numbered 25, what is the probability that Alicia will win on her next draw?

A. $\dfrac{1}{50}$

B. $\dfrac{1}{25}$

C. $\dfrac{2}{25}$

D. $\dfrac{4}{49}$

E. $\dfrac{1}{10}$

8. For all $x \neq 8$, $\dfrac{x^2 - 11x + 24}{8 - x} = ?$

F. $8 - x$

G. $3 - x$

H. $x - 3$

J. $x - 8$

K. $x - 11$

9. The ratio of girls to boys in a class is 3:5. If the total number of students is 32, how many more boys are there than girls?

A. 3

B. 5

C. 8

D. 12

E. 20

GO ON TO THE NEXT PAGE.

© 2006 Kaplan, Inc.

10. The table below displays Jamie's income for each of the years between 1989–1994. Which of the years between 1990–1994 show the greatest percent increase over the previous year?

DO YOUR FIGURING HERE.

Year	Income
1989	$20,000
1990	$25,000
1991	$30,000
1992	$33,000
1993	$36,000
1994	$44,000

F. 1990
G. 1991
H. 1992
J. 1993
K. 1994

11. Nancy and Meghan are playing a game that involves flipping a 2-sided coin designated heads and tails and rolling a 6-sided dice numbered 1 through 6. What is the probability of getting heads on the coin and an even number on the dice?

A. $\frac{1}{2}$

B. $\frac{1}{4}$

C. $\frac{1}{6}$

D. $\frac{1}{8}$

E. $\frac{1}{12}$

$\frac{1}{2} \times \frac{3}{6} = \frac{3}{12} = \frac{1}{4}$

GO ON TO THE NEXT PAGE.

© 2006 Kaplan, Inc.

12. Felicity earns $8.50 an hour for the first 40 hours she works each week. For each hour over 40 that she works, she earns $12.75. How many hours did she work in a week if she made $429.25?

- **F.** 35
- **G.** 42
- **H.** 45
- **J.** 47
- **K.** 52

DO YOUR FIGURING HERE.

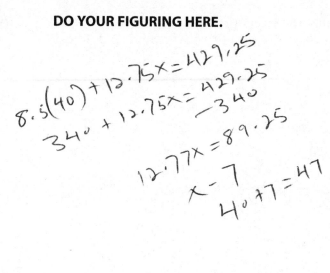

13. The following chart shows the type of pet that students in an 11th-grade class have. According to the graph, what percent of students do not own a pet?

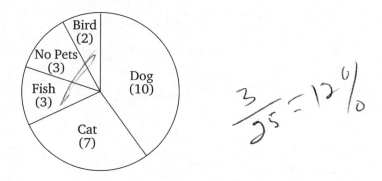

- **A.** 3%
- **B.** 8%
- **C.** 10%
- **D.** 12%
- **E.** 28%

14. Which of the following calculations will yield an odd integer for any even integer x?

- **F.** $x^3 + x$
- **G.** $x^3 + 1$
- **H.** $x + 4$
- **J.** $x^3 + x + 2$
- **K.** $x^3 + x^2 + 2$

GO ON TO THE NEXT PAGE.

© 2006 Kaplan, Inc.

15. Rachel's test scores were 79, 85, 63, 94, 79, 91, 85, 79, and 97. What was the mode of her test scores?

A. 63
B. 79
C. 85
D. 91
E. 97

END OF TEST.

STOP! DO NOT TURN THE PAGE UNTIL TOLD TO DO SO.

© 2006 Kaplan, Inc.

KAP Wrap

In this unit, you looked at Number Properties, Proportion/Probability, and how to read tables, graphs, and charts. Why is it so important to pay attention to all the information located within a graph or chart? Using information about the students in your class, create your own pie chart that you can then show your classmates.

© 2006 Kaplan, Inc.

Unit 4
Algebra

© 2006 Kaplan, Inc.

Thinking KAP

Your character, Oakthorn, finds a coded message in ancient runes, carved into a giant tree in the center of a forest. A good wizard has given you a book, with which you can decode the message.

Use the code book to translate the message below.

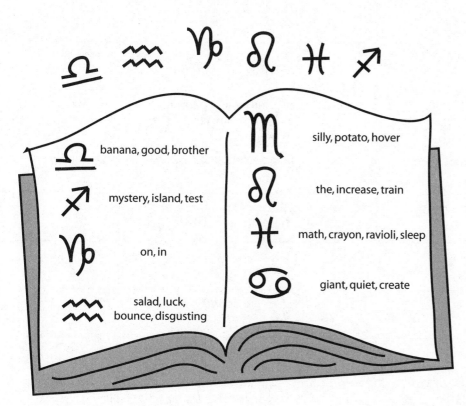

Strategy Instruction

Clues for Translation

Algebra is a language, just as English and Spanish are languages. Sometimes, before you solve a word problem, you need to translate that problem into an algebraic expression or equation. Clue words can help you do that. The chart below reviews useful clue words.

keep in mind

Mathematical terms can also help you translate a problem. For example, the clue word *sum* is a signal to add and the clue word *product* is a signal to multiply.

Clue Word	Symbol	In Words ...	In Algebraic Language ...
is/was/will be	=	The cost *is* $10.	$c = 10$
total	+	The *total* number of boys and girls was 14.	$b + g = 14$
plus	+	An amusement park charges $10 at the gate and $2 per ticket.	$10 + 2x$
more than	+	Maggie has $5 *more than* Kim.	$m = 5 + k$
less than	−	The height is 2 cm *less than* the length.	$h = l - 2$
of	×	75% *of* the regular cost	$0.75c$
per	×	$0.45 *per* pound	$0.45p$
twice	×	*twice* as tall as Ricky	$2r$
times	×	3 *times* as old as Darius	$3d$
half	÷ or /	*half* as many pumpkins	$\frac{1}{2}p$ or $\frac{p}{2}$

© 2006 Kaplan, Inc.

Use Logical Groups

When translating from words into algebra, translate in logical groups. Look for clue words to help to identify groups that can be translated into a word or symbol.

For each expression in words, find the mathematical expression that matches. Use clue words to help you. Let p stand for the number of pens Alice has.

_____ **1.** half as many pens as Alice

_____ **2.** twice as many pens as Alice

_____ **3.** 7 more pens than Alice

_____ **4.** 7 less than 3 times the number of pens Alice has

A. $2p$

B. $\dfrac{p}{2}$

C. $3p - 7$

D. $7 + p$

Try It Out!

Write your own expression for this problem. Then use that expression to solve the problem.

5. The cost of renting a car is $45 plus $0.10 for each mile traveled. How much would it cost to rent a car and drive it 50 miles?

- Underline the clue words.

- Write an expression for the cost of renting any car. Let m stand for the number of miles traveled.

- Substitute 50 for m in the expression. Find the answer.

keep in mind

Clue words are just guides to help you translate. Take the time to consider whether the algebraic expression or equation makes sense for the problem.

© 2006 Kaplan, Inc.

Picking Numbers

For test questions that ask you to identify the expression or equation for a particular problem, you can also use Picking Numbers.

Pick numbers to stand for the unknown values and use those numbers to solve the problem. Then substitute those same numbers into the answer choices to see which answer choices give the same result.

Picking Numbers
• Pick numbers to stand for unknown numbers or variables.
• Compute the answer using the number you picked.
• Substitute the value you picked into the expression in each answer choice. The expression that yields the same value as the one you found is the correct choice.

keep in mind

Whenever a problem has variables in the answer choices, think about whether or not Picking Numbers can help you.

Try It Out!

Use Picking Numbers to solve the problem below.

1. A bus holds 30 people, and a van holds 10 people. Which of the following expressions shows the number of people that could fill b buses and v vans?

    A. $b + v$

    B. $30b + 10v$

    C. $\dfrac{bv}{40}$

    D. $40bv$

    E. $10b + 30v$

© 2006 Kaplan, Inc.

Substitute and Compute

Sometimes you do not need to pick numbers to solve problems. Sometimes, a problem will tell you which numbers to pick. When a problem tells you the value of a variable and asks you to find its value, just Substitute and Compute.

Substitute and Compute
• Substitute a value for each variable.
• Use order of operations to simplify.

Here is a review of the correct order of operations.

1. First perform operations in _parentheses_.

2. Next, evaluate _exponents_.

3. _Multiply_ and _divide_ in order from left to right.

4. _Add_ and _subtract_ in order from left to right.

The memory device, _PEMDAS_, can help you recall the correct order of operations: _Parentheses, Exponents, Multiply and Divide, Add_ and _Subtract_.

keep in mind

Substitute and Compute can also help you solve function problems. For example, if $f(x) = 2x - 1$, then $f(3)$ can be found by substituting 3 for x in the expression $2x - 1$.

Try It Out!

Substitute and Compute to solve the problem below.

2.   If $x = 4$, then what is the value of $6x^2 - 3x^2 + 2x \cdot 5$?

Substitute 4 for x in the expression: $6x^2 - 3x^2 + 2x \cdot 5 = $ _____

Use the order of operations to simplify.

A

© 2006 Kaplan, Inc.

The 4-Step Method for Problem Solving

 STEP 1: **Understand the problem.**

STEP 2: **Analyze important information.**

STEP 3: **Plan and solve.**

STEP 4: **Check your work.**

Picking Numbers

- Pick numbers to stand for unknown numbers or variables.
- Compute the answer using the number you picked.
- Substitute the value you picked into the expression in each answer choice. The expression that yields the same value as the one you found is the correct choice.

Substitute and Compute

- Substitute a value for each variable.
- Use order of operations to simplify.

© 2006 Kaplan, Inc.

1. A salesperson earns $7h + 0.04s$ dollars, where h is the number of hours worked and s is the total amount of her sales. What does she earn for working 15 hours with a total of $120.50 in sales?

    A. $109.82

    B. $153.20

    C. $226.10

    D. $231.50

    E. $848.32

© 2006 Kaplan, Inc.

2. At a grocery store, item A costs $5 less than item B, and item B costs three times as much as item C. If n is the cost of item C in dollars, what is the cost of item A in dollars?

F. $n - 3$

G. $n - 5$

H. $3n - 5$

J. $3n - 15$

K. $5n$

© 2006 Kaplan, Inc.

A

Shared Practice

Use the 4-Step Method for Problem Solving and the strategies you learned in this lesson to solve the problems in this section.

1. If $n = 7$, then $2n + 3n + 4n + 5n + 6n$ is equivalent to:

   A. 70
   B. 84
   C. 98
   D. 126
   E. 140

   hint *Substitute and Compute. Be sure to follow the correct order of operations.*

2. Joan has q quarters, d dimes, n nickels, and no other coins in her purse. Which of the following represents the total number of coins in Joan's purse?

   F. $q + d + n$
   G. $5q + 2d + n$
   H. $0.25q + 0.10d + 0.05n$
   J. $(25 + 10 + 5)(q + d + n)$
   K. $25q + 10d + 5n$

   hint *Pay careful attention to what the question is asking. It is asking for the total number of coins, not the total value of those coins.*

© 2006 Kaplan, Inc.

3. A large jar holds 48 olives, and a small jar holds 32 olives. Which of the following expressions represents the number of olives needed to fill x large jars and y small jars?

    A. $48x + 32y$

    B. $\dfrac{x}{48} + \dfrac{y}{32}$

    C. $\dfrac{xy}{80}$

    D. $80xy$

    E. $\dfrac{1,536}{xy}$

    hint ▸ *Pick numbers for x and y. Solve the problem using those numbers. Then see which answer choice gives the same result.*

4. For a room, a hotel charges $55.90d + 35p$ dollars, where d is the number of days the room is rented and p is the number of people staying in the room. What would the hotel charge if a family of 3 people wanted to rent a room for 5 days?
    (Note: No sales tax is involved.)

    F. $98.90
    G. $202.70
    H. $342.70
    J. $384.50
    K. $727.20

    hint ▸ *Even though this is a word problem, all you really must do is Substitute and Compute.*

© 2006 Kaplan, Inc.

5. The total cost of renting a banquet hall is $102.00 for each hour the hall is used, plus $67.50 for each guest who attends. What is the total cost of renting the hall for a 4-hour event with 98 guests?
(Note: No sales tax is involved.)

    A. $ 678.00

    B $ 6,717.00

    C $7,023.00

    D $ 7,174.00

    E. $10,266.00

hint   *The clues "total" and "plus" tell you to add. The clue "each" tells you to multiply.*

6. If $f(x) = x^3 - x^2 - x$, what is the value of $f(^-3)$

    F. $^-39$

    G. $^-33$

    H. $^-21$

    J. $^-18$

    K.   0

hint   *To evaluate $f(^-3)$, just substitute $^-3$ for x in $x^3 - x^2 - x$. Then compute.*

© 2006 Kaplan, Inc.

7. Miguel is at a bookstore buying books, each of which usually costs $12.60. He learns that the books are on sale, and he can buy 5 books for a total of $55.00. If he buys the books on sale, and the sale price of each book is the same, by what amount is the price of *each* book reduced?

    A. $ 0.32

    B. $ 1.60

    C. $ 1.93

    D. $ 2.52

    E. $ 11.00

hint   *Use clue words to help you translate.*

8. The cost for a phone call is r cents for the first 3 minutes and s cents for each minute thereafter. What is the cost in cents of a phone call lasting exactly t minutes, where $t > 3$?

    F. $r + st$

    G. $r + s(t - 3)$

    H. $3r + st$

    J. $(3r + s)t$

    K. $\dfrac{3r + s}{t}$

hint   *Use Picking Numbers to help you solve the problem. Be sure to pick a number for t that is greater than 3.*

© 2006 Kaplan, Inc.

KAP Wrap

Look at the expression below.

$5a + 6$

Write a word problem that could be represented by this expression.
Include clue words in your problem to help whoever reads your problem
understand how to write an expression for it.

© 2006 Kaplan, Inc.

Equations and Inequalities

Thinking KAP

After an hour or so of playing video games, you realize you have spent $20.00. You can't believe the time has gone by so quickly! How many games have you played? You have no idea. You do know that the games cost $0.50 each, and you spent $1.50 on snacks. Can you figure out how many games you have played using this information? What strategies could you use to help you?

Solving Equations

One strategy you may have used to solve the problem in the Thinking KAP was to write and solve a **linear equation**. There are two ways to solve one-variable equations on the test: isolating the variable or Backsolving.

When Backsolving, if you try the middle answer choice and realize it is too great to be correct, then you can eliminate all the answer choices that are even greater. Doing this can save you valuable time on Test Day!

Backsolving
• Substitute each answer choice into the problem.   • Eliminate answer choices that do not satisfy the conditions of the problem.

Try It Out!

1. If $3x - 2 = 19$, then $x = $?

    A. $\dfrac{17}{3}$

    B. 7

    C. 8

    D. 17

    E. 21

$3x - 2 = 19$
$ +2 \quad +2$
_____
$\dfrac{3x}{3} = \dfrac{21}{3}$

$x = 7$

Use Backsolving. Try (C) and substitute 8 for x:

$3(8) - 2 \overset{?}{=} 19$

$ 22 \overset{?}{=} 19$

Is this statement true? _____

Backsolve with the remaining choices until you find the correct answer. What is the answer? _____

Now check your answer by isolating the variable and solving $3x - 2 = 19$ for x.

© 2006 Kaplan, Inc.

Systems of Equations

On the test, you may need to solve two equations considered together—a **system of equations**. To solve a system of equations, find the common solution for both equations. Backsolving can help you solve systems-of-equations problems on Test Day. However, you should also know how to use substitution.

Substitution

You can solve a system of equations algebraically using substitution.

Solve the system of equations: $^-x + y = {}^-2$ and $^-2x + y = 1$.

Rewrite in terms of y	Solve for x	Solve for y
Write the first equation in terms of y. $^-x + y = {}^-2$ $y = x - 2$	Substitute this value for y in the second equation. Solve for x. $^-2x + y = 1$ $^-2x + x - 2 = 1$ $^-x - 2 = 1$ $^-x = 3$ $x = {}^-3$	To find the value of y, substitute $^-3$ for x in either equation. $^-x + y = {}^-2$ $-(^-3) + y = {}^-2$ $3 + y = {}^-2$ $y = {}^-2 - 3 = {}^-5$

When using Backsolving to solve systems of equations, start with choice (A) and try the answer choices in order. Starting in the middle doesn't save time for questions involving systems of equations.

Try It Out!

2. What is the solution for x in this system of equations: $x + y = 6$ and $2x - y = 9$?

   F. 6
   G. 5
   H. 1
   J. $^-5$
   K. $^-6$

Use substitution to solve for x.

Now check your work with Backsolving.

$y = -x + 6$

$2x - 1(x + 6) = 9$

$2x + x - 6 = 9$

$ +6 \quad +6$

$\dfrac{3x}{3} = \dfrac{15}{3}$

$x = 5$

Quadratic Equations

A **quadratic equation** has a variable raised to the power of two. One way to solve a quadratic equation, such as $x^2 + 6x + 5 = 0$, is to factor.

Next, think about the last term. 5, factor pairs that yeild 5 include 1 and 5 or $^-1$ and $^-5$. Use the pair whose sum is the coefficient of the middle term of the polynomial, 6.

$$1 + 5 = 6$$

$$^-1 + {}^-5 = {}^-6$$

Use the factor pair 1 and 5.

$$(x + 1)(x + 5) = 0$$

You can check your work with FOIL (First-Inner-Outer-Last).

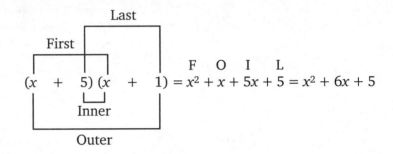

The equation $(x + 1)(x + 5) = 0$ would be true if either $(x + 1)$ or $(x + 5)$ were equal to zero. Solve for both cases.

$$(x + 1) = 0 \qquad\qquad (x + 5) = 0$$

$$x = {}^-1 \qquad\qquad\quad x = {}^-5$$

So the solutions are $^-1$ and $^-5$.

On a test you could also use backsolving

Try It Out!

Solve this problem by factoring or by Backsolving.

3. If $x^2 + 3x + 2 = 0$, what are the possible values for x?

    A. 1, 3
    B. 1, 2
    C. 0, $^-2$
    D. $^-1$, 2
    E. $^-1$, $^-2$

How could you use the strategy you didn't select to check your answer?

If zero is not on one side of a quadratic equation, manipulate the equation so that one side is zero before you factor. For example, to factor the equation $3x^2 + 10x + 6 = {}^-2$, add 2 to both sides so the equation becomes $3x^2 + 10x + 8 = 0$.

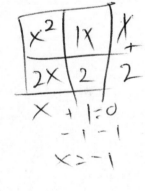

B

Matrix Multiplication

A **matrix** is a rectangular table of numbers.

The product of two matrices will have the same number of rows as the first matrix and the same number of columns as the second matrix. So, the product of the two matrices below will be a 2 by 1 matrix—that is, it will have two rows and one column.

In general, to find the entry for row x, column y of the product of two matrices, find row x in the first matrix and column y in the second matrix. Multiply each pair of corresponding entries and add their products. For example, to find the entry for Row 2, Column 1 of the matrices below, multiply the values in Row 2 of the first matrix by the corresponding values in Column 1 of the second matrix as shown below.

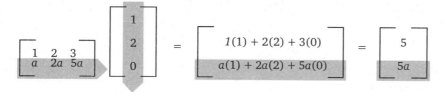

keep in mind

For test questions that ask for the matrix product, first find out how many rows and columns should be in the product. Eliminate answer choices with the wrong number of rows and columns.

The sum, $5a$, should be entered in Row 2, Column 1 of the matrix product, as shown above.

Try It Out!

4. What is the matrix product $\begin{bmatrix} {}^-6 & 2 \\ 0 & 1 \\ {}^-1 & {}^-2 \end{bmatrix} \begin{bmatrix} n \\ {}^-n \end{bmatrix}$?

First, decide how many rows and columns the matrix product should have.

• The first matrix has _____ row(s). The second matrix has _____ column(s).

• So, the product will have _____ row(s) and _____ column(s).

To find the entry for Row 1, Column 1, multiply the values in Row 1 of the first matrix by their corresponding entries in Column 1 of the second matrix. Fill in the entry for Row 1, Column 1 below.

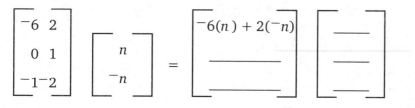

Complete the multiplication by finding the entry for Row 2, Column 1 and for Row 3, Column 1. Write the matrix product in the spaces above.

The 4-Step Method for Problem Solving

STEP 1: *Understand the problem.*

STEP 2: *Analyze important information.*

STEP 3: *Plan and solve.*

STEP 4: *Check your work.*

Backsolving
• Substitute each answer choice into the problem.
• Eliminate answer choices that do not satisfy the conditions of the problem.

© 2006 Kaplan, Inc.

1. Which of the following inequalities is equivalent to $^-6x - 2 \leq 10$?

    A. $x \leq ^-12$

    B. $x \leq ^-2$

    C. $x \geq 2$

    D. $x \geq ^-2$

    E. $x \geq ^-12$

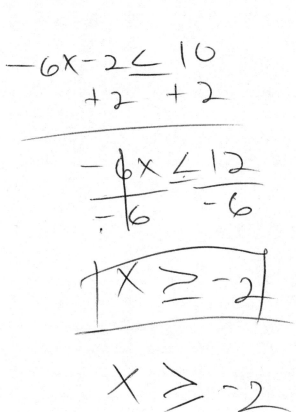

© 2006 Kaplan, Inc.

2. If $x^2 - 4x - 6 = 6$, what are the possible values for x?

    F.   4, 12

    G. $^-$6, 2

    H. $^-$6, $^-$2

    J.   6, 2

    K.   6, $^-$2

© 2006 Kaplan, Inc.

Shared Practice

Use the 4-Step Method for Problem Solving and the strategies you learned in this lesson to solve the problems in this section.

1. If $5x + 3 = 23$, then $x = ?$

   A. $^-25$

   B. $^-7$

   C. $^-4$

   D. $\dfrac{1}{3}$

   E. 4

   hint *Start with middle choice (C) and Backsolve. If (C) is too great to be correct, eliminate the even greater numbers. If (C) is too small to be correct, eliminate the even smaller numbers.*

2. If $x - 2y = 0$, and $3x + y = 7$, then what is the value of x?

   F. $^-1$

   G. 0

   H. 1

   J. 2

   K. 3

   hint *Try Backsolving. Start with (H) and substitute the numbers in the answer choices for x. Find the value of x that gives the same value of y in both equations.*

B

© 2006 Kaplan, Inc.

3.   The formula for converting a Fahrenheit temperature reading to Celsius is

$$C = \frac{5}{9}(F - 32)$$

where C is the reading in degrees Celsius and F is the reading in degrees Fahrenheit. Which of the following is the Fahrenheit equivalent to a reading of 95° Celsius?

A.   35° F

B.   53° F

C.   63° F

D.   203° F

E.   207° F

hint *This problem is really about solving an equation. Substitute 95 for C. Use Backsolving to find the answer choice that gives that value of C.*

4.   If $f(x) = x^2 + 3$, then $f(x - k) = ?$

F. $x^2 - k^2$

G. $x^2 - 3 - k^2$

H. $x^2 + k^2 + 3$

J. $x^2 - 2xk + k^2$

K. $x^2 - 2xk + k^2 + 3$

hint *Substitute x - k for x in the expression $x^2 + 3$. Use the FOIL method to evaluate the expression.*

ACT ADVANTAGE
MATHEMATICS

© 2006 Kaplan, Inc.

5. Which of the following is the solution statement for the inequality $^-3 < 4x - 5$?

  A. $x > {}^-2$

  B. $x > \dfrac{1}{2}$

  C. $x < {}^-2$

  D. $x < \dfrac{1}{2}$

  E. $x > 2$

  hint *Do not automatically reverse the inequality symbol. Only do so if you multiply or divide by a negative number.*

6. Find the matrix product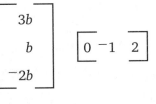

  F. $\begin{bmatrix} 0 & 0 & 0 \\ ^-3b & ^-b & 2b \\ 2b & 2b & ^-4b \end{bmatrix}$

  G. $\begin{bmatrix} 0 & ^-3b & 6b \\ 0 & ^-b & 2b \\ 0 & 2b & ^-4b \end{bmatrix}$

  H. $\begin{bmatrix} 0 & ^-b & ^-4b \end{bmatrix}$

  J. $\begin{bmatrix} 0 & ^-2b & 4b \end{bmatrix}$

  K. $\begin{bmatrix} ^-5b \end{bmatrix}$

  hint *How many rows and columns should the product have? Which choices can you eliminate just based on that?*

© 2006 Kaplan, Inc.

7. If $k - 3 = \frac{5}{3}$, then $3 - k = ?$

    A. $-\frac{5}{3}$

    B. $-\frac{3}{5}$

    C. $\frac{3}{5}$

    D. $\frac{5}{3}$

    E. $\frac{7}{3}$

hint *Rewrite the first equation in terms of k so that positive 3 is on one side of the equation. Keep manipulating the equation until 3 - k is on one side and a number is on the other.*

8. What is the sum of the values of x for which $2x^2 = 2x + 12$?

    F. $^-2$

    G. $^-1$

    H. 0

    J. 1

    K. 2

hint *Since the question asks for the sum of the values of x, you cannot Backsolve. Rewrite the equation so 0 is one side of the equal sign, and factor.*

B

© 2006 Kaplan, Inc.

KAP Wrap

Imagine that a friend was absent on the day Backsolving was taught in class. Choose a problem in the Shared Practice which could be solved by Backsolving. On the lines below, write a step-by-step explanation for your friend of how Backsolving could be applied to that problem.

© 2006 Kaplan, Inc.

Working with Patterns

ReKAP

Review the strategies from Lessons A and B. Then fill in the blanks with what you have learned.

1. Two of the clue words that are often used to signal addition are _____ and _____.

2. The memory device _____ can help me recall the correct order of operations to use when I Substitute and Compute.

3. When I use Backsolving, I substitute the numbers in the answer choices for the given variable until I find the number that _____ _____.

4. If I know that the product of two matrices will be a 2 by 3 matrix, I can eliminate any answer choices that do not have _____ rows and _____ columns.

Strategy Instruction

Patterns and Sequences

Some questions on the test will involve patterns and sequences. Many test questions will state what kind of sequence is involved. Because of this, it is worth your time to learn the difference.

keep in mind

In an arithmetic sequence, the difference between pairs of consecutive numbers is always the same. For example, in the sequence 4, 6, 8, 10, ... the difference between consecutive numbers is always 2.

Type of Pattern/ Sequence	How Terms Are Found	Examples
arithmetic sequence	Each new number is obtained from the previous number by adding or subtracting a constant.	4, 6, 8, 10, ... (add 2) 26, 21, 16, 11, ... (subtract 5)
geometric sequence	Each new number is obtained from the previous number by multiplying by a constant.	5, 25, 125, 625, ...(multiply by 5) 64, 32, 16, 8, ... (multiply by .5)
repeating pattern	A pattern with a string of terms that repeats.	5, 6, 6, 5, 6, 6, ... (5, 6, 6, repeats) blue, red, blue, red, ... (terms blue and red alternate)

 *Try It Out!*

Answer each question. Explain how you determined each answer.

1. Look at this arithmetic sequence: 18, 16, 14, 12, ... What is the missing number in this sequence? ____14____

   (handwritten: –2 –2 –2)

2. Look at this geometric sequence: 5, 10, 20, 40, ... What is the next number in this sequence? ____80____

   (handwritten: +5 +10 +20)

3. Erneshaya's little brother made a pattern out of shapes. The pattern repeated: square, circle, circle, rectangle, square, circle, circle, rectangle, ... What will be the 10th shape in this pattern? ___Circle___

   (handwritten: Square, Circle, Circle.)

© 2006 Kaplan, Inc.

Make a Chart

Not all test questions will ask you to identify the next number in a sequence. They may instead ask you to do something more complex, such as identifying the 100th term in a sequence. The sequence may involve variables. Making a chart can help you solve more complex problems.

Try It Out!

Use the chart below to help you solve this problem.

1. The first, second, and third terms in a geometric sequence are $3x$, $9x$, and $27x$, in that order. What is the 100th term in that sequence?

    A. $3x^{100}$
    B. $9x^{100}$
    C. $3^{99}x$
    D. $3^{100}x$
    E. $9^{99}x$

keep in mind

If you are unsure of how to solve a problem, look at the answer choices. They may provide a clue to help you organize the information you are given in a helpful way.

Position	Term
1	$3x = 3^1x$
2	$9x = 3^2x$
3	$27x = 3^3x$
...	...
100	$3^{100}x$

Fill in the chart. Since the problem involves exponents, think about how you can rewrite each term with an exponent. The first one has been filled in for you.

How can you use the chart to determine which answer choice is correct?

_____

_____

Name_____ Date_____

Test Practice Unit 4

When your teacher tells you, carefully tear out this page. Then begin working.

1. Ⓐ Ⓑ Ⓒ Ⓓ Ⓔ   11. Ⓐ Ⓑ Ⓒ Ⓓ Ⓔ

2. Ⓕ Ⓖ Ⓗ Ⓙ Ⓚ   12. Ⓕ Ⓖ Ⓗ Ⓙ Ⓚ

3. Ⓐ Ⓑ Ⓒ Ⓓ Ⓔ   13. Ⓐ Ⓑ Ⓒ Ⓓ Ⓔ

4. Ⓕ Ⓖ Ⓗ Ⓙ Ⓚ   14. Ⓕ Ⓖ Ⓗ Ⓙ Ⓚ

5. Ⓐ Ⓑ Ⓒ Ⓓ Ⓔ   15. Ⓐ Ⓑ Ⓒ Ⓓ Ⓔ

6. Ⓕ Ⓖ Ⓗ Ⓙ Ⓚ

7. Ⓐ Ⓑ Ⓒ Ⓓ Ⓔ

8. Ⓕ Ⓖ Ⓗ Ⓙ Ⓚ

9. Ⓐ Ⓑ Ⓒ Ⓓ Ⓔ

10. Ⓕ Ⓖ Ⓗ Ⓙ Ⓚ

© 2006 Kaplan, Inc.

MATHEMATICS TEST

15 Minutes—15 Questions

DIRECTIONS: Solve each problem, choose the correct answer, and then fill in the corresponding oval on your answer document.

Do not linger over problems that take too much time. Solve as many as you can; then return to the others in the time you have left for this test.

You are permitted to use a calculator on this test. You may use your calculator for any problems you choose, but some of the problems may best be done without using a calculator.

Note: Unless otherwise stated, all of the following should be assumed.

1. Illustrations are NOT necessarily drawn to scale.
2. Geometric figures lie in a plane.
3. The word *line* indicates a straight line.
4. The word *average* indicates arithmetic mean.

---

1. If $x = {}^-2$, then $14 - 3(x + 3) = ?$

    A. $^-1$

    B. 11

    C. 14

    D. 17

    E. 29

DO YOUR FIGURING HERE

2. To complete a certain task, Group A requires 8 more hours than Group B, and Group B requires twice as long as Group C. If h is the number of hours required by Group C, how long does the task take Group A, in terms of h?

    F. $10h$

    G. $16h$

    H. $10 + h$

    J. $2(8 + h)$

    K. $8 + 2h$

© 2006 Kaplan, Inc.

3. Mr. Miller is buying turkey at the deli. The turkey usually costs $6.20 per pound, but today the turkey is on sale. Mr. Miller buys 3 pounds of turkey on sale for a total of $17.55. By what amount is the price per pound of turkey reduced?

A. $ 0.35

B. $ 1.05

C. $ 2.83

D. $ 5.85

E. $11.35

4. If $a(x) = \sqrt{x^2} + 7$ and $b(x) = x^3 - 7$, then what is the value of $\dfrac{a(3)}{b(2)}$?

F. $\dfrac{\sqrt{11}}{20}$

G. $\dfrac{1}{5}$

H. $\sqrt{11}$

J. 10

K. $4\sqrt{11}$

5. If $47 - x = 188$, then $x = $?

A. $^-235$

B. $^-141$

C. 4

D. 141

E. 235

GO ON TO THE NEXT PAGE.

ACT ADVANTAGE
MATHEMATICS

© 2006 Kaplan, Inc.

6. What is the solution for x in this system of equations?

$$^-3x + y = 1$$
$$^-x + y = {}^-3$$

F. 5

G. 2

H. 1

J. $^-2$

K. $^-5$

7. Which of the following is the solution statement for the inequality $3x - 2 > 10$?

A. $x > {}^-4$

B. $x > 4$

C. $x > 12$

D. $x < 4$

E. $x < 12$

8. If $x = t + 2$ and $y = 4 - t$, which of the following expresses y in terms of x?

F . $y = 2 - \dfrac{1}{3}x$

G. $y = 6 - x$

H. $y = 2 - 3x$

J. $y = -2 - 3x$

K. $y = 6 - 3x$

9. If $x \neq 0$, and $x^2 - 3x = 6x$, then $x = $?

A. $^-9$

B. $^-3$

C. $\sqrt{3}$

D. 3

E. 9

GO ON TO THE NEXT PAGE.

© 2006 Kaplan, Inc.

10. The following matrix shows the number of each type of sandwich purchased for a teacher's meeting.

$$\begin{array}{cccc} \text{Tuna} & \text{Turkey} & \text{Cheese} & \text{Ham} \\ \begin{bmatrix} 12 & 15 & 10 & 8 \end{bmatrix} \end{array}$$

The matrix below shows the price of each of these types of sandwiches, including tax.

$$\begin{array}{cc} \text{Tuna} & \\ \text{Turkey} & \\ \text{Cheese} & \\ \text{Ham} & \end{array} \begin{bmatrix} \$5.15 \\ \$5.25 \\ \$3.95 \\ \$5.00 \end{bmatrix}$$

Given these matrices, what was the total cost of the sandwiches ordered for the meeting?

F. $217.69
G. $219.05
H. $220.05
J. $231.75
K. $232.20

11. A gumball machine dispenses gumballs of different colors in the following pattern: blue, red, red, yellow, white, white, green, green, green. If the pattern keeps repeating, what will be the color of the 89th gumball?

A. Blue
B. Red
C. Yellow
D. White
E. Green

12. In the geometric sequence 3, 12, r, 192, ... what is the value of r, the third term?

F. 24
G. 36
H. 48
J. 60
K. 96

GO ON TO THE NEXT PAGE.

© 2006 Kaplan, Inc.

13. What 3 numbers should be placed in the blanks below so that the difference between consecutive numbers is the same?

12, ___ , ___ , ___ , 32

- **A.** 16, 22, 28
- **B.** 17, 22, 27
- **C.** 20, 22, 30
- **D.** 22, 24, 30
- **E.** 23, 29, 31

14. The first and second terms of a geometric sequence are 2*n* and 4*n*, in that order. What is the 500th term in the sequence?

- **F.** $2n^{499}$
- **G.** $4n^{299}$
- **H.** $2^{499}n$
- **J.** $2^{500}n$
- **K.** $4^{499}n$

DO YOUR FIGURING HERE

GO ON TO THE NEXT PAGE.

© 2006 Kaplan, Inc.

15. The pattern below shows how toothpicks can be used to make squares. Which of the following describes the total number of toothpicks used to make *n* squares in the arrangement illustrated below?

DO YOUR FIGURING HERE

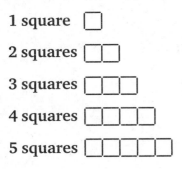

1 square

2 squares

3 squares

4 squares

5 squares

- **A.** The total number of toothpicks is always equal to 13 toothpicks, regardless of the number of squares.

- **B.** The total number of toothpicks is always twice the number of squares.

- **C.** The total number of toothpicks is always 1 more than 3 times the number of squares.

- **D.** The total number of toothpicks is always 4 times the number of squares.

- **E.** There is no consistent relationship between this total and the number of squares.

END OF TEST.

STOP! DO NOT TURN THE PAGE UNTIL TOLD TO DO SO.

© 2006 Kaplan, Inc.

KAP Wrap

Create an arithmetic sequence. Write the first five terms of your sequence below.

_____

How do you know that the sequence you wrote is an arithmetic sequence and not a geometric sequence?

© 2006 Kaplan, Inc.

Unit 5
Geometry

© 2006 Kaplan, Inc.

Rules for Polygons

Thinking KAP

In the fantasy video game you are playing, the castle is being destroyed, and you are stranded on the square roof of a tower nearby. Fortunately, you have special powers that enable you to create a circular force field to protect yourself against enemies and flying debris. You know that the area of the square roof is 36 square feet. How can you calculate the largest possible force field that you will be able to make, without going past the edge of the roof?

Strategy Instruction

Quadrilaterals

A polygon with four sides is called a **quadrilateral**. There are five quadrilaterals that appear frequently on the ACT.

keep in mind

A trapezoid has two different-sized bases. To find the area of a trapezoid, multiply the average of the two bases by the height.

Type	Characteristics	Type	Characteristics
Square	• All sides are the same length. • Opposite sides are parallel. • Four right angles	Parallelogram	• Opposite sides are the same length. • Opposite sides are parallel. • Opposite angles are equal.
Rectangle	• Opposite sides are the same length. • Opposite sides are parallel. • Four right angles	Rhombus	• All sides are the same length. • Opposite sides are parallel. • Opposite angles are equal.
Trapezoid	• Only two sides are parallel.		

There are several aspects of a quadrilateral that can be measured or calculated. The most common ones are perimeter and area.

- **Perimeter:** Find the total distance around the figure by adding all the side lengths.

- **Area:** Find the total area inside the figure by mulitplying the base by the height.

© 2006 Kaplan, Inc.

Circles

A circle is not a polygon because it is not made of line segments. Know the following circle formulas for Test Day.

- **Circumference:** $C = 2\pi r$

- **Area:** $A = \pi r^2$

To find the area of a sector of a circle, divide the degree measure of the central angle by 360°, and multiply the result by the area of the circle.

Try It Out!

1. What is the area, in square inches, of a circle with a circumference of 2π inches?

    A. $\dfrac{\pi}{2}$

    B. π

    C. 2π

    D. 4π

    E. 8π

keep in mind

The circumference is a measure of the distance around the edge of a circle, just as the perimeter of a polygon is the distance around the edge of a figure.

© 2006 Kaplan, Inc.

UNIT 5: GEOMETRY
LESSON A: RULES FOR POLYGONS

193

Triangles

Triangles are three-sided polygons. They may be characterized by their angle measures or side lengths.

Type	Example	Characteristics
Types of Triangles by Angle Measure		
Acute Triangle		All angles measure less than 90°.
Right Triangle		One angle measures exactly 90°.
Obtuse Triangle		One angle measures greater than 90°.
Types of Triangles by Side Length		
Equilateral Triangle		All sides have the same length, and all angles have the same measure.
Isosceles Triangle		Two sides have the same length, and the two angles formed by sides of different lengths have the same measure.
Scalene Triangle		No sides have the same length, and no angles have the same measure.

Congruent polygons are identical—they have the same size and shape. Similar polygons have equal angle measures but only proportional sides.

The **area** of a triangle is one-half of the product of its base and height.

$$A = \frac{1}{2}bh.$$

In all triangles, the sum of the three interior angles is 180°, and the sum of the two shortest sides is greater than the length of the longest side.

© 2006 Kaplan, Inc.

Right Triangles

Right triangles have one angle measuring 90°. The two sides that form the 90° angle are the *legs* of the triangle, and the third side is called the *hypotenuse*.

The lengths of the legs and hypotenuse are proportional, as expressed in the Pythagorean theorem. The Pythagorean theorem relates the side lengths of a right triangle by the formula $a^2 + b^2 = c^2$, where a and b are the lengths of the legs and c is the length of the hypotenuse.

There are special combinations of side lengths in which all of the values are integers. These are called Pythagorean triples. The most common triples are $3 : 4 : 5$ and $5 : 12 : 13$.

There are two special types of right triangles. A *scalene right triangle* has no equal side lengths. There is a special type of scalene right triangle that has angles measuring 30°, 60°, and 90°. An *isosceles right triangle* has angles that measure 45°, 45°, and 90°, and its legs are identical in length.

keep in mind

If a figure is not given to you in a problem, it is often useful to draw a picture to help you visualize the situation.

Scalene Right Triangle

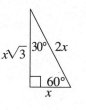

Isosceles Right Triangle

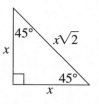

Try It Out!

2. A ladder that is 20 feet long is placed against a wall such that the base of the ladder is 12 feet from the wall. How many feet above the ground is the top of the ladder?

    F.   8

    G.   12

    H.   15

    J.   16

    K.   18

© 2006 Kaplan, Inc.

The 4-Step Method for Problem Solving

STEP 1: *Understand the problem.*

STEP 2: *Analyze important information.*

STEP 3: *Plan and solve.*

STEP 4: *Check your work.*

© 2006 Kaplan, Inc.

1. In parallelogram *WXYZ*, shown below, the perimeter is 30 units. How many units long is line segment $\overline{XY}$?

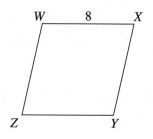

A. $\dfrac{3}{\frac{3}{4}}$

B.   7

C.   8

D.   11

E.   12

© 2006 Kaplan, Inc.

2. In circle O below, points R, O, and T are colinear. If angle SOT measures 30°, what is the measure of angle SRT?

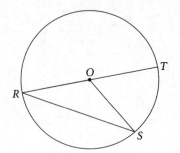

    F.   10°

    G.   15°

    H.   20°

    J.   25°

    K.   30°

© 2006 Kaplan, Inc.

Shared Practice

Use the 4-Step Method for Problem Solving and the strategies you have learned in this lesson to solve the problems in this section.

1. Find the perimeter of the following triangle.

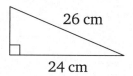

26 cm

24 cm

- **A.** 10 cm
- **B.** 50 cm
- **C.** 60 cm
- **D.** 120 cm
- **E.** 624 cm

> **hint** ▷ *The perimeter of a polygon is the distance around the figure.*

2. What is the area of the parallelogram below?

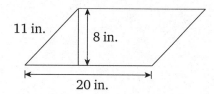

11 in.

8 in.

20 in.

- **F.** 28 in.^{2}
- **G.** 39 in.^{2}
- **H.** 88 in.^{2}
- **J.** 160 in.^{2}
- **K.** 220 in.^{2}

> **hint** ▷ *The formula for the area of a parallelogram is $A = b \times h$.*

© 2006 Kaplan, Inc.

3. What is the circumference of the circle below?

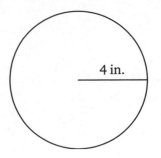

4 in.

    A. 4π in.

    B. 8π in.

    C. 14π in.

    D. 16π in.^{2}

    E. 18π in.^{2}

> **hint**   *Use the formula for the circumference of a circle.*

4. Triangle *ABC*, shown below, is equilateral, with side lengths $4\sqrt{2}$ units. What is the length of line segment $\overline{BD}$?

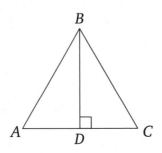

    F. $3\sqrt{6}$ units

    G. $2\sqrt{6}$ units

    H. $3\sqrt{3}$ units

    J. 4 units

    K. $2\sqrt{2}$ units

> **hint**   *If you know $\overline{AC}$, you can find $\overline{AD}$. Then you can use the Pythagorean theorem.*

© 2006 Kaplan, Inc.

5. In the figure below, what is the length of line segment *AD*?

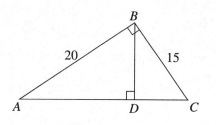

A. 9 units

B. 12 units

C. 15 units

D. 16 units

E. 25 units

hint ▷ *Use the formula that relates the side lengths of right triangles.*

6. How many feet long is the diameter of a circle that has a circumference of 36π feet?

F. 12

G. 18

H. 6π

J. 36

K. 18π

hint ▷ *Determine how diameter and radius are related, and then use the formula for the circumference of a circle to solve.*

© 2006 Kaplan, Inc.

7. In isosceles triangle *ABC*, line segments $\overline{AB}$ and $\overline{BC}$ are congruent. If $\angle CAB = 27°$, what is the measure of $\angle ABC$?

   A.   27°

   B.   54°

   C.   90°

   D.   126°

   E.   153°

   hint *Draw a picture and use the properties of an isosceles triangle to find the unknown angle measure.*

8. In order to triple the area of a square, the new side lengths must be equal to the old side lengths multiplied by:

   F. $\sqrt{3}$

   G.   3

   H.   9

   J. $\sqrt{27}$

   K.   27

   hint *Use Backsolving to solve this problem.*

© 2006 Kaplan, Inc.

KAP Wrap

In this lesson, you learned about the perimeter of polygons, the circumference of circles, and the area of polygons and circles. List three real-world situations that would require the use of one of these three features of polygons or circles.

© 2006 Kaplan, Inc.

Complex Figures

Thinking KAP

In the fantasy video game you are playing, you find a potion that will protect your character, Oakthorn, from arrows, slingshots, and large animals. This special potion is called Vaniloakpickloranchocorosedragoscale. Unfortunately you are not sure exactly what ingredients are needed to make this strange brew. Use the information in the name of the potion to determine the ingredients that are probably needed.

© 2006 Kaplan, Inc.

Areas of Complex Polygons

To calculate the area of a complex figure, begin by looking for squares, rectangles, and triangles, and drawing lines to show the simpler figures. Then find the area of each simple figure and add the areas of all of the parts together to find the entire area of the complex figure.

Occasionally, a problem may require you to find the area of a shaded section of a figure. For this type of problem, it is sometimes easier to find the area of the unshaded section, and subtract it from the area of the entire figure.

To find the area of a complex figure, break the figure down into the simplest possible shapes.

Try It Out!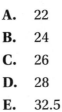

1. What is the area, in square inches, of the trapezoid below?

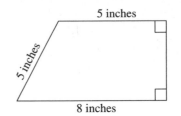

5 inches

5 inches

8 inches

    A. 22

    B. 24

    C. 26

    D. 28

    E. 32.5

© 2006 Kaplan, Inc.

Areas of Complex Figures Involving Circles

When one figure is placed inside another, it is said to be **inscribed**. When a circle is placed around a figure so that it is tangent at each vertex, the circle is said to **circumscribe** the figure.

To solve problems with inscribed or circumscribed figures, combine rules for each figure present.

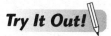

Try It Out!

2. The figure below shows square *OABC* and circle *O*. If points *A* and *C* lie on the circumference of the circle, and the area of the square is 16 square units, what is the area of the circle?

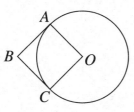

keep in mind

> Once you know one measure of a circle, you can find most others.

- **F.** 4 square units
- **G.** 4π square units
- **H.** 8π square units
- **J.** 32 square units
- **K.** 16π square units

© 2006 Kaplan, Inc.

B

Finding the Perimeters of Irregular Polygons

As you learned in Lesson A, the perimeter of any polygon can be calculated by adding the lengths of its sides. However, finding the perimeter of an irregular polygon can sometimes be more challenging than finding the perimeter of a regular polygon.

When possible, use your knowledge of parallel and perpendicular lines in such problems.

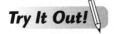
Try It Out!

As with regular polygons, the perimeters of irregular polygons can be found by adding the lengths of the different parts that form their boundaries.

3. All angles in the figure below are right angles. What is the perimeter of the figure?

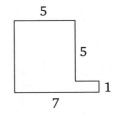

  A. 18

  B. 20

  C. 24

  D. 26

  E. 35

© 2006 Kaplan, Inc.

Angles, Lines, and Line Segments

When two lines intersect, a variety of angle types may be formed.

Types of Angles	What are they?	What else should you know about them?
vertical angles	opposite angles formed at the intersection of two lines	Vertical angles have the same degree measure.
supplementary angles	adjacent angles sharing a line	Supplementary angles sum to 180 degrees.
corresponding angles	angles formed in the same position when parallel lines are intersected	Corresponding angles have equal degree measures.

If a line or line segment divides an angle or line segment into two equal parts, it **bisects** it. When a transversal crosses two parallel lines, the angles created have important relationships.

keep in mind

Any angle, line, or line segment that is bisected is divided into two identical pieces.

Try It Out!

What angle relationships do you see in the diagram below?

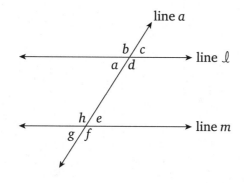

The 4-Step Method for Problem Solving

STEP 1: **Understand the problem.**

STEP 2: **Analyze important information.**

STEP 3: **Plan and solve.**

STEP 4: **Check your work.**

© 2006 Kaplan, Inc.

1. In the figure below, if segment $\overline{AC}$ is parallel to segment $\overline{ED}$, what is the length of segment $\overline{AE}$?

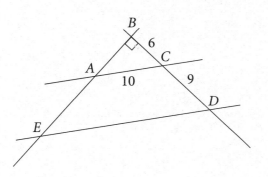

A. 8

B. 10

C. 12

D. 15

E. 21

© 2006 Kaplan, Inc.

2. The trapezoid below is divided into two triangles and one rectangle. What is the combined area, in square inches, of the two triangles?

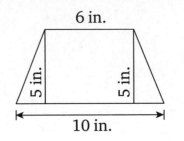

F. 10 in.^{2}

G. 36 in.^{2}

H. 40 in.^{2}

J. 50 in.^{2}

K. 60 in.^{2}

© 2006 Kaplan, Inc.

Shared Practice

Use the 4-Step Method for Problem Solving and the strategies you learned in this lesson to solve the problems in this section.

1.   In the figure below, line *m* is parallel to line *n*, point *A* lies on line *m*, and points *B* and *C* lie on line *n*. What is the measure of ∠*y*?

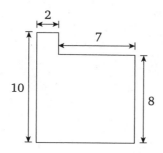

   A.   44°

   B.   90°

   C.   124°

   D.   136°

   E.   154°

   hint   *Remember the characteristics of angles and lines.*

2.   If all angles in the figure below are right angles, what is the perimeter of the figure?

   F.   18

   G.   27

   H.   38

   J.   76

   K.   90

   hint   *Opposite sides have the same length in parallelograms and rectangles.*

© 2006 Kaplan, Inc.

3.   The front surface of a fence panel is shown below. The panel is symmetrical along its center vertical axis. What is the area, in square inches, of the front surface of the panel?

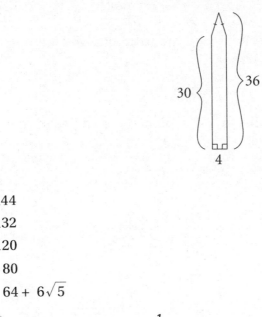

A.   144
B.   132
C.   120
D.   80
E. $64 + 6\sqrt{5}$

hint *The area of a triangle is $\frac{1}{2}bh$.*

4.   The circle in the figure below is inscribed in a square that has a perimeter of 16 inches. What is the area, in square inches, of the shaded region?

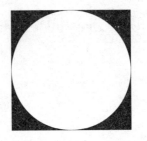

F. 4π
G. $16 - 2\pi$
H. $16 - 4\pi$
J. $8 - 2\pi$
K. $8 - 4\pi$

hint *Think of the area of the square and the area of the circle separately.*

© 2006 Kaplan, Inc.

5. What is the measure of ∠*ABC* in the figure below?

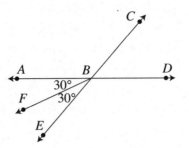

- **A.** 30°
- **B.** 60°
- **C.** 90°
- **D.** 120°
- **E.** It cannot be determined from the information given.

hint ▷ *Supplementary angles can be made up of more than two angles.*

6. If all angles in the figure below are right angles, what is the perimeter of the figure?

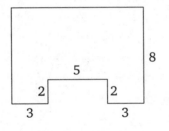

- **F.** 19
- **G.** 21
- **H.** 27
- **J.** 38
- **K.** 42

hint ▷ *Opposite sides will have equal values.*

B

© 2006 Kaplan, Inc.

7. What is the area, in square inches, of the rectangle in the figure below?

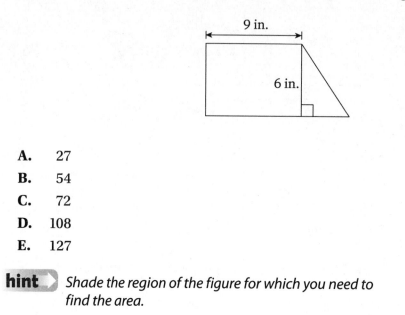

A. 27

B. 54

C. 72

D. 108

E. 127

hint ▷ *Shade the region of the figure for which you need to find the area.*

8. In the figure below, circle *R* is tangent to circle *S*. Point *S* lies on the circumference of circle *R*. If the area of circle *R* is 6 square inches, what is the area, in square inches, of circle *S*?

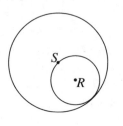

F. 12

G. 24

H. 36

J. 12π

K. 36π

hint ▷ *Look at the placement of Circle R for information about the radius of Circle S.*

© 2006 Kaplan, Inc.

KAP Wrap

Where might you encounter complex figures in the real world? Why is it important to know how to find their areas or to break them down into smaller, simpler figures?

© 2006 Kaplan, Inc.

B

ReKAP

Review the strategies from Lessons A and B. Then fill in the blanks to show what you have learned.

1. The _____ is the distance around the outside of an object.

2. The amount of space inside an object is called the _____.

3. The formula for the circumference of a circle is _____.

4. The formula for the area of a triangle is _____.

© 2006 Kaplan, Inc.

Strategy Instruction

Solid Figures and Volume

In Lessons A and B, you learned characteristics of two-dimensional figures. In this lesson, you will learn characteristics of three-dimensional figures.

One special measurement for solids is volume, the amount of space an object occupies. Below is a table that shows how to calculate volume for the solids that are most common on the ACT.

keep in mind

Volume is measured in cubic units (cm^3, $in.^3$, ft^3) because it is three-dimensional.

Solid	Example	Volume Formula
Cube		$V = s^3$
Rectangular Prism		$V = \ell wh$
Cylinder		$V = \pi r^2 h$

Try It Out!

1. Each side of a cube measures 5 inches. What is the volume of the cube?

    A. 125 in.^{3}

    B. 30 in.^{3}

    C. 25 in.^{3}

    D. 15 in.^{3}

    E. 10 in.^{3}

© 2006 Kaplan, Inc.

Surface Area

Surface area is the total area of an object's shell, or outside. Imagine peeling off the outer layer of an object. The area of that layer is the object's surface area. Therefore, the surface area of a three-dimensional object is the sum of the areas of the sections on its exterior. The surface area formulas for the most common three-dimensional objects on the ACT are given below.

Solid	Example	Surface Area Formula
Cube		$SA = 6s^2$
Rectangular Prism		$SA = 2\ell w + 2\ell h + 2wh$
Cylinder		$SA = 2\pi r^2 + 2\pi rh$

keep in mind

Like area, surface area is measured in square units (cm^2, $in.^2$, ft^2).

Try It Out!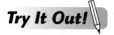

2. Each side of a cube measures 5 inches. What is the surface area, in square inches, of this cube?

    F.   25 in.^{2}

    G.   30 in.^{2}

    H.   125 in.^{2}

    J.   150 in.^{2}

    K.   200 in.^{2}

© 2006 Kaplan, Inc.

Test Practice Unit 5

When your teacher tells you, carefully tear out this page. Then begin working.

1. Ⓐ Ⓑ Ⓒ Ⓓ Ⓔ   11. Ⓐ Ⓑ Ⓒ Ⓓ Ⓔ

2. Ⓕ Ⓖ Ⓗ Ⓙ Ⓚ   12. Ⓕ Ⓖ Ⓗ Ⓙ Ⓚ

3. Ⓐ Ⓑ Ⓒ Ⓓ Ⓔ   13. Ⓐ Ⓑ Ⓒ Ⓓ Ⓔ

4. Ⓕ Ⓖ Ⓗ Ⓙ Ⓚ   14. Ⓕ Ⓖ Ⓗ Ⓙ Ⓚ

5. Ⓐ Ⓑ Ⓒ Ⓓ Ⓔ   15. Ⓐ Ⓑ Ⓒ Ⓓ Ⓔ

6. Ⓕ Ⓖ Ⓗ Ⓙ Ⓚ

7. Ⓐ Ⓑ Ⓒ Ⓓ Ⓔ

8. Ⓕ Ⓖ Ⓗ Ⓙ Ⓚ

9. Ⓐ Ⓑ Ⓒ Ⓓ Ⓔ

10. Ⓕ Ⓖ Ⓗ Ⓙ Ⓚ

© 2006 Kaplan, Inc.

6. In circle *O* shown below, *P*, *O*, and *Q* are colinear. If ∠*ROQ* measures 50°, what is the measure of ∠*RPQ*?

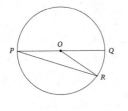

  F. 20°

  G. 25°

  H. 30°

  J. 35°

  K. 40°

7. A triangle has sides of lengths 6 units, 8 units, and 10 units, and a square has a perimeter of 28 units. What is the positive difference, in square units, between the area of the triangle and the area of the square?

  A.   1

  B.   4

  C. 25

  D. 49

  E. 148

8. The circle below has a diameter of 6 inches. What is the length, in inches, of the arc that has a central angle of 60°?

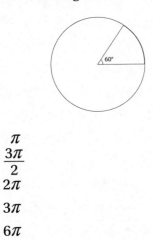

  F. π

  G. $\dfrac{3\pi}{2}$

  H. 2π

  J. 3π

  K. 6π

GO ON TO THE NEXT PAGE.

© 2006 Kaplan, Inc.

9. What is the area, in square units, of the figure below?

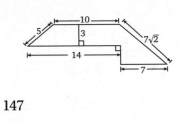

- **A.** 147
- **B.** 108.5
- **C.** 91
- **D.** 60.5
- **E.** $39 + 7\sqrt{2}$

10. In the figure below, the area of parallelogram *PQRS* is 24 square units. Segment *QT* is perpendicular to segment *PS*, and point *T* is the midpoint of segment *PS*. What is the perimeter, in units, of *PQRS*?

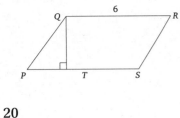

- **F.** 20
- **G.** 22
- **H.** 24
- **J.** 26
- **K.** 28

11. How many units long is the radius of a circle that has a circumference of 24π?

- **A.** 6
- **B.** 12
- **C.** 24
- **D.** 12π
- **E.** 24π

GO ON TO THE NEXT PAGE.

© 2006 Kaplan, Inc.

12. A right triangle has sides measuring 5 inches, 12 inches, and 13 inches. What is the area, in square inches, of this triangle?

DO YOUR FIGURING HERE.

 F. 12

 G. 15

 H. 30

 J. 60

 K. 65

13. A circle has a diameter of 8 in. What is the area, in square inches, of this circle?

 A. 4

 B. 8π

 C. 16

 D. 16π

 E. 64π

14. What is the volume of a rectangular prism with a length of 8 in., a width of 7 in., and a height of 6 in.?

 F. 21 in.^{2}

 G. 48 in.^{2}

 H. 56 in.^{2}

 J. 256 in.^{3}

 K. 336 in.^{3}

15. The diagonal of a square has a length of 8 cm. What is the area of the square?

 A. $4\sqrt{2}$ cm^2

 B. $8\sqrt{2}$ cm^2

 C. 16 cm^2

 D. 32 cm^2

 E. 64 cm^2

END OF TEST.

STOP! DO NOT TURN THE PAGE UNTIL TOLD TO DO SO.

KAP Wrap

In this unit, you looked at characteristics of two- and three-dimensional figures. Compare and contrast the characteristics of two- and three-dimensional figures. You may wish to make a chart to show the differences and similarities between them. Then, give two examples of two- and three-dimensional figures that you encounter every day in the real world.

© 2006 Kaplan, Inc.

Unit 6
Coordinate Geometry and Trigonometry

© 2006 Kaplan, Inc.

Thinking KAP

As Oakthorn, the elf hero, you appear to be lost in the forest, but you must get to the castle to save the queen. The only things you have with you are a map and a compass. What is the easiest way to get to the castle? What is the fastest way to get to the castle?

Coordinate Geometry

Points, lines, and curves can all be plotted on the coordinate plane. In order to describe where these things are located on the plane, mathematicians use a series of "road signs."

You can remember that in an ordered pair, the *x*-value comes before the *y*-value, just as *x* comes before *y* in the alphabet.

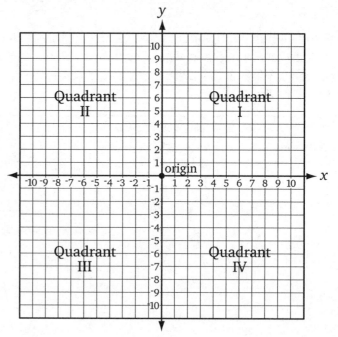

The values of *x* and *y* that describe a specific point are known as **coordinates**. The first number tells you how far to move left and right on the *x*-axis. The second number tells you how far to move up and down on the *y*-axis. The *x*- and *y*-values of the ordered pair can be positive or negative, depending upon the quadrant in which the plotted point is located. Complete the table below to show the signs of the *x* and *y* when the plotted point is located in a given quadrant.

Quadrant	(x, y)
I	(+, +)
II	(,)
III	(,)
IV	(,)

© 2006 Kaplan, Inc.

Draw It

When you are given a coordinate geometry problem without a coordinate plane, begin by drawing your own plane and plotting the given coordinate points.

Try It Out!

1. In the standard (x, y) coordinate plane, the points $(2, 2)$, $(2, {}^-3)$, and $({}^-4, 2)$ represent three corners of a rectangle. What are the coordinates of the rectangle's fourth corner?

   A. $({}^-4, 2)$

   B. $({}^-4, {}^-3)$

   C. $({}^-4, 3)$

   D. $({}^-3, {}^-4)$

   E. $(4, 3)$

Use the space below to sketch a coordinate plane. Then plot the given points on the graph.

Don't confuse the coordinates. The x-coordinate always comes first.

© 2006 Kaplan, Inc.

Finding the Length of a Line Segment

Finding the distance of horizontal and vertical lines is straightforward. Just subtract the coordinates that are different.

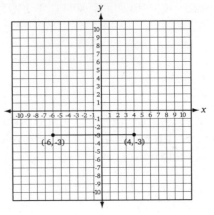

In the example above, the line segment is horizontal, and the coordinates of the endpoints are (4, ⁻3) and (⁻6, ⁻3). The y-coordinates of the endpoints are the same: ⁻3. So find the difference between 4 and ⁻6:

$$4 - (^-6) = 10$$

To find the length of a sloping line segment, draw a right triangle and use the Pythagorean theorem. Draw lines parallel to the x-axis and y-axis to create a right triangle in which the given line segment is a hypotenuse. Find the lengths of the legs, substitute these values into the Pythagorean formula, and solve.

If you have trouble subtracting signed numbers, use a picture to count the total distance from one endpoint to another.

Try It Out!

2. What is the length of the line segment in the graph below?

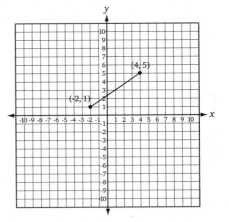

A. 3 units
B. 5 units
C. $\sqrt{47}$ units
D. 7 units
E. $2\sqrt{13}$ units

Draw a right triangle so that the line segment you want to measure forms the hypotenuse. Then use the Pythagorean theorem.

© 2006 Kaplan, Inc.

Finding the Midpoint of a Line Segment

The midpoint is the point that divides a segment into two equal parts. If you know the endpoints of a line segment, you can find its midpoint by using the midpoint formula.

Midpoint Formula $\dfrac{x_1 + x_2}{2}, \dfrac{y_1 + y_2}{2}$

This formula seems complicated, but all it really means is that, to find the midpoint, find the average (mean) of the x-coordinates and the average (mean) of the y-coordinates.

Look at the graph below.

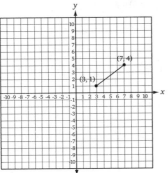

keep in mind

Don't be thrown by complicated formulas. Just think about what the formula means.

To find the x-coordinate of the midpoint of this line segment, average the x-coordinates.

$$\frac{7 + 3}{2}$$

To find the y-coordinate of the midpoint of this line segment, average the y-coordinates. The midpoint is (5, _____).

Try It Out!

3. What is the midpoint of the line segment below?

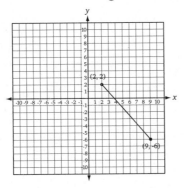

 A. $(7, {}^{-}3)$

 B. $(5.5, {}^{-}4)$

 C. $(5, 0)$

 D. $(5.5, {}^{-}2,)$

 E. $({}^{-}3, 7)$

Plot the ordered pair on the graph. Does your answer seem reasonable?

© 2006 Kaplan, Inc.

The 4-Step Method for Problem Solving

STEP 1: *Understand the problem.*

STEP 2: *Analyze important information.*

STEP 3: *Plan and solve.*

STEP 4: *Check your work.*

© 2006 Kaplan, Inc.

1. What is the length of a line segment with endpoints at ($^-$2, 6) and (3, $^-$6)?

   A.　　1 unit

   B.　　5 units

   C.　　10 units

   D.　　13 units

   E.　　17 units

© 2006 Kaplan, Inc.

2. What is the length of $\overline{AB}$ in the diagram below?

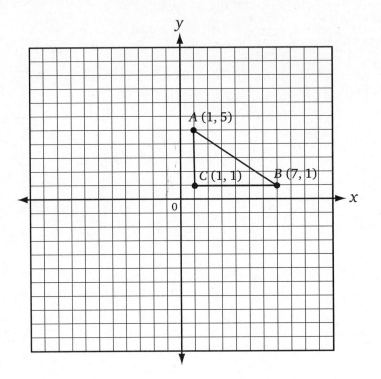

F. 52 units

G. 12 units

H. $2\sqrt{13}$ units

J. 4 units

K. $\sqrt{13}$ units

© 2006 Kaplan, Inc.

Shared Practice

Use the 4-Step Method for Problem Solving and the strategies you have learned in this lesson to solve the problems in this section.

1. Points A (4, 7) and B (2, 1) determine $\overline{AB}$ in the standard (x, y) coordinate plane.

   Find the midpoint of $\overline{AB}$.

   A. (4, 3)
   B. (6, 8)
   C. (⁻3, 1)
   D. (11, 3)
   E. (3, 4)

   hint *Find each coordinate separately by averaging.*

2. Which of the following points lies on the line $y = \frac{1}{2}x + 3$?

   F. (0, 0)
   G. (3, 0)
   H. (0, 3)
   J. (6, 4.5)
   K. (6.5, 4)

   hint *Draw a picture to solve this problem.*

© 2006 Kaplan, Inc.

3.   Which point on the number line below represents $(0, {}^-5)$?

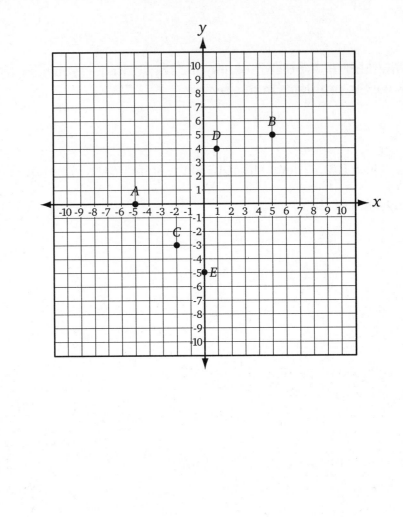

   A.   *A*
   B.   *B*
   C.   *C*
   D.   *D*
   E.   *E*

> **hint** ▸ *Remember which value is plotted first.*

4.   What is the distance between the points $({}^-3, {}^-2)$ and $({}^-5, 3)$?

   F.   2 units
   G. $\sqrt{23}$ units
   H. $\sqrt{29}$ units
   J.   5 units
   K. $\sqrt{33}$ units

> **hint** ▸ *Draw a right triangle.*

© 2006 Kaplan, Inc.

5. A straight line in the coordinate plane passes through the points with (x, y) coordinates (5, 2) and (‾3, 0). What are the (x, y) coordinates of the point at which the line passes through the y-axis?

A. $(0, \frac{4}{5})$

B. $(0, \frac{3}{4})$

C. $(0, 0)$

D. $(0, -\frac{3}{4})$

E. $(0, ‾3)$

hint *Draw the line to help find the answer.*

6. In the standard (x, y) coordinate plane, at which x-value does the line described by $3x + y = 9$ intersect the x-axis?

F. ‾9

G. ‾3

H. 0

J. 3

K. 9

hint *This problem can be solved algebraically to find x.*

A

© 2006 Kaplan, Inc.

7. In the standard (x, y) coordinate plane, if the distance between $(a, 2)$ and $(16, a)$ is 10 units, which of the following could be the value of a?

    A.   10

    B.   6

    C.   ⁻6

    D.   ⁻8

    E.   ⁻10

    hint Solve this using the distance formula and substituting in values.

8. In the standard (x, y) coordinate plane, points A and B lie on line ℓ, and point C is on the origin. The coordinates of points A and B are $(1, 5)$ and $(5, 1)$. What is the shortest distance from point C to line ℓ?

    F. $2\sqrt{3}$ units

    G.   4 units

    H. $3\sqrt{2}$ units

    J. $3\sqrt{3}$ units

    K.   5 units

    hint Draw a picture to help you visualize what's going on.

© 2006 Kaplan, Inc.

KAP Wrap

In this lesson, you analyzed line segments. On the blank coordinate grid below, draw a line segment. Then describe the segment with as many details as you can, using the strategies you learned in this lesson. Be sure to include the length of the segment, the midpoint, and the coordinates of the endpoints.

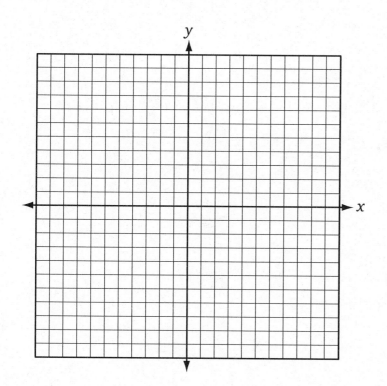

Graphing Equations

Thinking KAP

As Oakthorn, the elf hero, you have successfully made your way to the castle. Unfortunately, the entrance is locked and you have no way in. As you stop to think of ways to enter the castle, you look up and see hundreds of keys hanging right above the door. However, you realize that by the time you try all of the keys, it might be too late to save the queen. You must come up with a method of finding the correct key without trying all of the keys. You notice that the keys come in all different sizes, shapes, and weights. How can you use this information to find which key is the correct key?

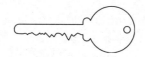

© 2006 Kaplan, Inc.

Graphing Linear Equations

You already know how to plot points on a coordinate plane. Using that knowledge, you can graph linear equations as well.

Three important values can be found from a linear equation: the slope, the *x*-intercept, and the *y*-intercept. These values are easy to identify when the equation is written in slope-intercept form:

$$y = mx + b$$

where *m* is the slope and *b* is the *y*-intercept. For each value of *x* in a linear equation, there is one value of *y*, and vice versa.

keep in mind

Slopes of parallel lines are identical, while slopes of perpendicular lines are negative reciprocals of each other.

Term	Definition	How To Find It
Slope	the ratio of the change in *y* over the change in *x* (this describes the rise over run, or the slant of the line)	1. Pick two points on the line. 2. Find the difference between the *y*- and *x*-coordinates and place them in a ratio. $m = \dfrac{y_2 - y_1}{x_2 - x_1}$
x-intercept	the point at which the line crosses the *x*-axis	1. Set *y* = 0 in the equation. 2. Solve for *x*. The resulting value is the *x*-intercept.
y-intercept	the point at which the line crosses the *y*-axis	1. Set *x* = 0 in the equation. 2. Solve for *y*. The value found is the *y*-intercept.

B

© 2006 Kaplan, Inc.

Parallel and Perpendicular Lines

Parallel lines are two straight lines that never intersect. Because parallel lines have the same slope, the *m*-value of two parallel lines in the form $y = mx + b$ will always be the same.

Perpendicular lines are two straight lines that meet at a right angle. Their slopes are negative reciprocals. So in the equation $y = mx + b$, if the slope of one line is 3, then the slope of a line perpendicular to it will be $-\frac{1}{3}$.

In the graph below, line *m* and line *n* are parallel. Line *o* is perpendicular to both lines.

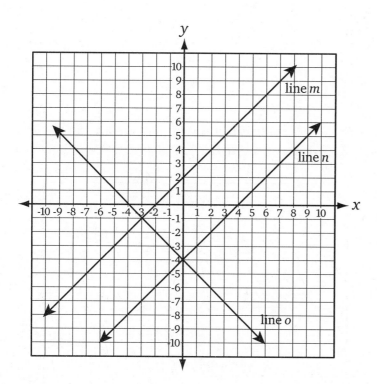

keep in mind

Lines that are perpendicular to the same line are parallel.

Try It Out!

1. What is the slope of a line that is perpendicular to the line $y = \frac{1}{2}x + 5$?

   A. $^-6$

   B. $^-2$

   C. 2

   D. 5

   E. 6

B

Eliminating Incorrect Answers

When given a graph and asked to find the equation that corresponds to the graph, or vice versa, look for ways to eliminate incorrect answers.

Consider the type of equation and the *y*-intercept point. Eliminate answers that do not fit.

Type of Equation	Form of Equation	Example	*y*-intercept
Linear (line)	$y = mx + b$	$y = 3x + 4$	4
Quadratic (parabola)	$y = ax^2 + bx + c$	$y = x^2 + 5x + 6$	6

keep in mind

In a quadratic equation, if the parabola opens up, the *a*-value is positive, and if it opens down, the *a*-value is negative.

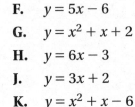

Try It Out!

2. Which of the following is the equation of the line on the graph below?

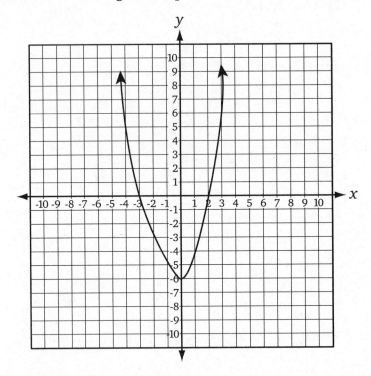

   F. $y = 5x - 6$

   G. $y = x^2 + x + 2$

   H. $y = 6x - 3$

   J. $y = 3x + 2$

   K. $y = x^2 + x - 6$

B

© 2006 Kaplan, Inc.

Backsolving In Coordinate Geometry

Some coordinate geometry problems are most effectively solved by working backwards. Among these are problems that:

- give an equation, and ask you to select the graph of that equation.

- show the graph of a line and ask you to identify the equation of the line.

- give a set of points and require you to identify an equation that passes through them.

In these cases, you can substitute numbers from the problem into the answer choices to find the one that works.

Backsolving

- Put each answer choice back into the problem. If an answer choice does not make the statement true, eliminate it.

keep in mind

Graphing can take a long time on an exam. Look for shortcuts to eliminate answers, or solve through Backsolving to use your time more efficiently.

Try It Out!

3.   When graphed on the coordinate plane, which of the following equations passes through the points ($^-$2, 4) and (1, 4) ?

   A. $y = x^2$
   B. $y = x^2 + 2x + 1$
   C. $y = 2x^2 + 2x$
   D. $y = 5x^2 - 1$
   E. $y = 6x^2 + x + 1$

Use Backsolving to solve this problem.

- Substitute the numbers from the first coordinate into each choice. Which choice(s) contain that point?

- Now substitute the numbers from the second coordinate into the remaining choices. Eliminate choices until only one remains.

B

© 2006 Kaplan, Inc.

The 4-Step Method for Problem Solving

 STEP 1: *Understand the problem.*

STEP 2: *Analyze important information.*

STEP 3: *Plan and solve.*

STEP 4: *Check your work.*

Backsolving
• Put each answer choice back into the problem. If an answer choice does not make the statement true, eliminate it.

© 2006 Kaplan, Inc.

1. Line x is perpendicular to line y, and line y is perpendicular to line z. If the slope of line x is 4, what is the slope of line z?

   A. $^-4$

   B. $-\dfrac{1}{4}$

   C. 1

   D. 4

   E. It cannot be determined from the information provided.

© 2006 Kaplan, Inc.

2. In the standard (x, y) coordinate plane, what is the x-intercept of the line $3x + y = 9$?

    F. $^-9$

    G. $^-3$

    H. $-\dfrac{1}{3}$

    J. $\quad3$

    K. $\quad9$

B

© 2006 Kaplan, Inc.

Shared Practice

Use the 4-Step Method for Problem Solving and the strategies you have learned in this lesson to solve the problems in this section.

1. What is the slope of any line parallel to the line $-\frac{1}{6}x + \frac{2}{3}y = 12$?

  A. $\frac{1}{4}$

  B. $\frac{2}{3}$

  C. 1

  D. 4

  E. 8

> **hint** *Remember the special characteristic of parallel slopes, and be sure to put the given equation in slope-intercept form.*

2. Two lines, A and B, are perpendicular. If the equation of line A is $y = \frac{2}{3}x + 6$, what could be the equation of line B?

  F. $y = 3x + 7$

  G. $y = -\frac{3}{2}x - 4$

  H. $y = -\frac{2}{3}x + 6$

  J. $y = {}^{-}2x + 14$

  K. $y = {}^{-}3x + 2$

> **hint** *Think about the relationship between the slopes of perpendicular lines.*

B

© 2006 Kaplan, Inc.

3.   What is the *y*-intercept of the line $20x + 4y = 40$?

    A.   ⁻5

    B.   10

    C.   4

    D.   ⁻10

    E.   1

 Substitute 0 for x to find the y-intercept. Check your work by sketching the line.

4.   What is the slope of the line that passes through the origin and (7, 5)?

    F. 5

    G. 7

    H. $\dfrac{7}{5}$

    J. $\dfrac{5}{7}$

    K. 2

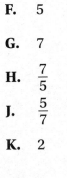 *Remember that the origin is the point (0, 0).*

© 2006 Kaplan, Inc.

B

5. Line t in the standard (x, y) coordinate plane has a y-intercept of $^-3$ and is parallel to the line having the equation $3x - 5y = 4$. Which of the following is an equation for line t?

A. $y = -\dfrac{3}{5}x + 3$

B. $y = -\dfrac{5}{3}x - 3$

C. $y = \dfrac{3}{5}x + 3$

D. $y = \dfrac{5}{3}x + 3$

E. $y = \dfrac{3}{5}x - 3$

hint *Think about the slopes of parallel lines.*

6. In the standard (x, y) coordinate plane, line l is perpendicular to the line containing the points $(5, 6)$ and $(6, 10)$. What is the slope of line l?

F. $^-4$

G. $-\dfrac{1}{4}$

H. 0

J. 4

K. 8

hint *Think about the slopes of perpendicular lines.*

B

7. The line that passes through the points (1, 1) and (2, 16) in the standard (x, y) coordinate plane is parallel to the line that passes through the points ($^-$10, $^-$5) and (a, 25). What is the value of a?

    A. $^-$15

    B. $^-$10

    C. $^-$8

    D. $^-$4

    E. 10

    hint Parallel lines have the same slope. Set slopes equal to each other to solve.

8. What is the equation of the graph shown below?

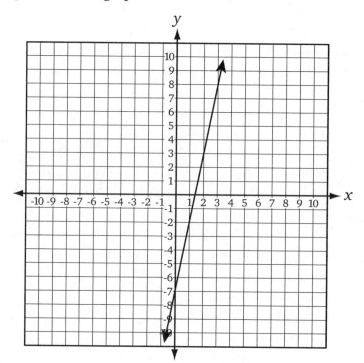

    F. $y = 5x^2 - 7x + 1$

    G. $y = 5x - 7$

    H. $y = x^2 + 5x - 7$

    J. $y = 3x - 5$

    K. $y = 5x + 7$

    hint Solve using Eliminating and Backsolving.

© 2006 Kaplan, Inc.

KAP Wrap

Explain how using the answer choices can help you earn points on ACT
Coordinate Geometry questions.

B

© 2006 Kaplan, Inc.

lesson C

Trigonometry

ReKAP

Review the strategies from Lessons A and B. Then fill in the blanks with what you have learned.

1. The _____ is the vertical number line in the coordinate plane.

2. The length of a line segment can be found by using the _____.

3. I can find the _____ of a line segment by computing the average of the two *x*-values of the endpoints and the average of the two *y*-values of the endpoints and then rewriting the two resulting numbers as an ordered pair.

4. To find the slope of a line, I can look at the coefficient of *x* when a linear equation is written in _____.

© 2006 Kaplan, Inc.

Trigonometry

Six trigonometric functions are tested on the ACT: sine (sin), cosine (cos), tangent (tan), secant (sec), cotangent (cot), and cosecant (csc).

These functions relate either of the two non-right angles of a right triangle to lengths of sides of the triangle. The angles are often identified by a Greek letter, most often θ (pronounced "theta"), α (pronounced "alpha"), or β (pronounced "beta"). Be sure to label the opposite and adjacent sides correctly, based on which angle is selected. (Note that the hypotenuse is always the side opposite the right angle.)

keep in mind

You can remember to which side of a right triangle each trig function relates by using the acronym SOHCAHTOA.

$$\sin\theta = \frac{\mathbf{o}\text{pposite}}{\mathbf{h}\text{ypotenuse}}$$

$$\cos\theta = \frac{\mathbf{a}\text{djacent}}{\mathbf{h}\text{ypotenuse}}$$

$$\tan\theta = \frac{\sin\theta}{\cos\theta} = \frac{\mathbf{o}\text{pposite}}{\mathbf{a}\text{djacent}}$$

$$\sec\theta = \frac{1}{\cos\theta} = \frac{\text{hypotenuse}}{\text{adjacent}}$$

$$\cot\theta = \frac{1}{\tan\theta} = \frac{\text{adjacent}}{\text{opposite}}$$

$$\csc\theta = \frac{1}{\sin\theta} = \frac{\text{hypotenuse}}{\text{opposite}}$$

If you know the measure of the angle you are using as your reference angle, and you have a graphing calculator, you can check your answer to any of these functions. Make sure the mode on your calculator is set to degrees, not radians. Let's say that your solution to a trigonometry problem is $\cos 60° = \frac{1}{2}$. You can type "cos 60" into your calculator to confirm that it equals $\frac{1}{2}$.

© 2006 Kaplan, Inc.

Trigonometry Shortcuts

Memorize the following two shortcuts. They can save you time on Test Day.

$$\sin^2\theta + \cos^2\theta = 1$$

$$\frac{\sin\theta}{\cos\theta} = \tan\theta$$

Try It Out!

1. $\dfrac{\sin^2\theta + \cos^2\theta}{\cos\theta} =$

   A. $\tan\theta$

   B. $\mathrm{scs}\theta$

   C. $\cos\theta$

   D. $\sin\theta$

   E. $\sec\theta$

Use a shortcut to rewrite the numerator.

keep in mind

Checking the answer choices can help you identify the kind of the answer you need.

Does the result look similar to one of the values described on the previous page?

© 2006 Kaplan, Inc.

Test Practice Unit 6

When your teacher tells you, carefully tear out this page. Then begin working.

1. Ⓐ Ⓑ Ⓒ Ⓓ Ⓔ   11. Ⓐ Ⓑ Ⓒ Ⓓ Ⓔ

2. Ⓕ Ⓖ Ⓗ Ⓙ Ⓚ   12. Ⓕ Ⓖ Ⓗ Ⓙ Ⓚ

3. Ⓐ Ⓑ Ⓒ Ⓓ Ⓔ   13. Ⓐ Ⓑ Ⓒ Ⓓ Ⓔ

4. Ⓕ Ⓖ Ⓗ Ⓙ Ⓚ   14. Ⓕ Ⓖ Ⓗ Ⓙ Ⓚ

5. Ⓐ Ⓑ Ⓒ Ⓓ Ⓔ   15. Ⓐ Ⓑ Ⓒ Ⓓ Ⓔ

6. Ⓕ Ⓖ Ⓗ Ⓙ Ⓚ

7. Ⓐ Ⓑ Ⓒ Ⓓ Ⓔ

8. Ⓕ Ⓖ Ⓗ Ⓙ Ⓚ

9. Ⓐ Ⓑ Ⓒ Ⓓ Ⓔ

10. Ⓕ Ⓖ Ⓗ Ⓙ Ⓚ

© 2006 Kaplan, Inc.

MATHEMATICS TEST

15 Minutes—15 Questions

DIRECTIONS: Solve each problem, choose the correct answer, and then fill in the corresponding oval on your answer document.

Do not linger over problems that take too much time. Solve as many as you can; then return to the others in the time you have left for this test.

You are permitted to use a calculator on this test. You may use your calculator for any problems you choose, but some of the problems may best be done without using a calculator.

Note: Unless otherwise stated, all of the following should be assumed.

1. Illustrations are NOT necessarily drawn to scale.
2. Geometric figures lie in a plane.
3. The word *line* indicates a straight line.
4. The word *average* indicates arithmetic mean.

---

1. What are the coordinates of point *A* in the coordinate plane below?

DO YOUR FIGURING HERE.

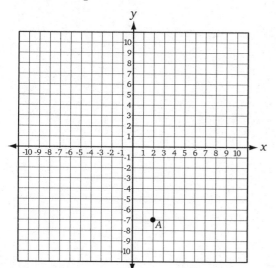

   A.   (⁻7, 2)

   B.   (2, 7)

   C.   (5, ⁻2)

   D.   (⁻5, ⁻5)

   E.   (2, ⁻7)

2. What is the length of a line segment with endpoints at (⁻4, ⁻4) and (7, ⁻4)?

   F.   3

   G.   4

   H.   11

   J.   12

   K.   13

GO ON TO THE NEXT PAGE.

© 2006 Kaplan, Inc.

3. In the line below, what is the midpoint of the line segment $\overline{AB}$?

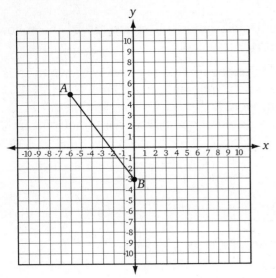

- **A.** $(^-3, 1)$
- **B.** $(0, 2)$
- **C.** $(1, 4)$
- **D.** $(3, 4)$
- **E.** $(6, 2)$

4. What is the distance length of the line segment with endpoints $(5, 4)$ and $(^-1, ^-1)$?

- **F.** 6
- **G.** $\sqrt{52}$
- **H.** $\sqrt{61}$
- **J.** 8
- **K.** $\sqrt{87}$

5. Given a line with the equation $y = ^-\dfrac{7}{8} x + \dfrac{1}{4}$, what is the slope of a line that is perpendicular to this line?

- **A.** $\dfrac{7}{8}$
- **B.** $\dfrac{8}{7}$
- **C.** $-\dfrac{8}{7}$
- **D.** $\dfrac{1}{4}$
- **E.** $-\dfrac{1}{8}$

GO ON TO THE NEXT PAGE.

© 2006 Kaplan, Inc.

6. What is the slope of the line with the equation $-3x + 10y = 18$?

DO YOUR FIGURING HERE.

    F. $\dfrac{3}{10}$

    G. $\dfrac{9}{5}$

    H. 3

    J. $\dfrac{10}{3}$

    K. 18

7. What are the coordinates of the x-intercept in the equation $y = 5x + 6$?

    A. $(0, 1)$

    B. $(0, -\dfrac{5}{6})$

    C. $(6, 0)$

    D. $(5, 0)$

    E. $(-\dfrac{6}{5}, 0)$

8. What could be the equation of the line in the following graph?

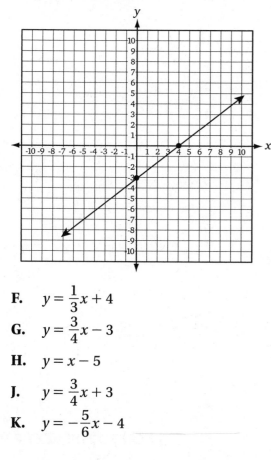

    F. $y = \dfrac{1}{3}x + 4$

    G. $y = \dfrac{3}{4}x - 3$

    H. $y = x - 5$

    J. $y = \dfrac{3}{4}x + 3$

    K. $y = -\dfrac{5}{6}x - 4$

GO ON TO THE NEXT PAGE.

© 2006 Kaplan, Inc.

9. Two lines are perpendicular. What is the slope of the second line if the equation of the first line is $^-6x + \frac{1}{5}y = 3$?

A. $\frac{6}{5}$

B. 15

C. $-\frac{1}{30}$

D. 30

E. $-\frac{1}{15}$

DO YOUR FIGURING HERE.

10. What is the midpoint of a line segment with endpoints at $(\frac{1}{2}, {}^-3)$ and $(\frac{3}{8}, 6)$?

F. $(\frac{7}{16}, \frac{3}{2})$

G. $(\frac{7}{8}, 3)$

H. $(\frac{2}{5}, 9)$

J. $(3, -\frac{1}{2})$

K. $(5, 2)$

11. If two of the points on a line are $(^-2, 5)$ and $(4, {}^-6)$, what is its slope?

A. $^-2$

B. $-\frac{1}{2}$

C. $\frac{1}{2}$

D. $\frac{6}{11}$

E. $-\frac{11}{6}$

GO ON TO THE NEXT PAGE.

12. Triangle *ABC* is shown below. What is the length of line segment *BC*?

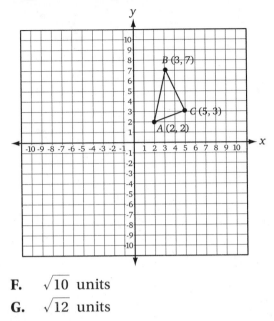

F. $\sqrt{10}$ units

G. $\sqrt{12}$ units

H. 4 units

J. 6 units

K. $\sqrt{20}$ units

13. What is the equation of the following graph?

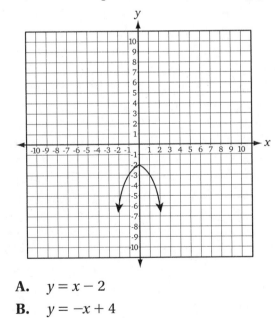

A. $y = x - 2$

B. $y = -x + 4$

C. $y = -x^2 - 2$

D. $y = x^2 + 4$

E. $y = x^2 - 2$

GO ON TO THE NEXT PAGE.

© 2006 Kaplan, Inc.

14. What is the slope of the line $y = \square x + \triangle$?

   F. y

   G. $\triangle$

   H. $\dfrac{\square}{\triangle}$

   J. $\square$

   K. x

15. In the figure below, PQR is a right triangle, and p, q, and r represent the lengths, in units, of the sides of the triangle. What is the cosecant of angle RQP?

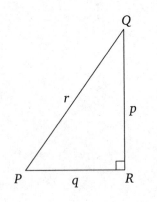

   A. $\dfrac{p}{q}$

   B. $\dfrac{p}{r}$

   C. $\dfrac{q}{p}$

   D. $\dfrac{q}{r}$

   E. $\dfrac{r}{q}$

END OF TEST.

STOP! DO NOT TURN THE PAGE UNTIL TOLD TO DO SO.

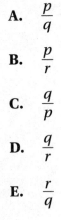

© 2006 Kaplan, Inc.

KAP Wrap

In this unit, you looked at coordinate geometry and trigonometry. Why is it important to know if a triangle is a right triangle? How can knowing that information help you when finding the length of a line segment or when solving a trigonometric problem?

© 2006 Kaplan, Inc.

Practice Test 2

© 2006 Kaplan, Inc.

Practice Test 2

When your teacher tells you, carefully tear out this page. Then begin working.

1. Ⓐ Ⓑ Ⓒ Ⓓ Ⓔ   15. Ⓐ Ⓑ Ⓒ Ⓓ Ⓔ   29. Ⓐ Ⓑ Ⓒ Ⓓ Ⓔ

2. Ⓕ Ⓖ Ⓗ Ⓙ Ⓚ   16. Ⓕ Ⓖ Ⓗ Ⓙ Ⓚ   30. Ⓕ Ⓖ Ⓗ Ⓙ Ⓚ

3. Ⓐ Ⓑ Ⓒ Ⓓ Ⓔ   17. Ⓐ Ⓑ Ⓒ Ⓓ Ⓔ   31. Ⓐ Ⓑ Ⓒ Ⓓ Ⓔ

4. Ⓕ Ⓖ Ⓗ Ⓙ Ⓚ   18. Ⓕ Ⓖ Ⓗ Ⓙ Ⓚ   32. Ⓕ Ⓖ Ⓗ Ⓙ Ⓚ

5. Ⓐ Ⓑ Ⓒ Ⓓ Ⓔ   19. Ⓐ Ⓑ Ⓒ Ⓓ Ⓔ   33. Ⓐ Ⓑ Ⓒ Ⓓ Ⓔ

6. Ⓕ Ⓖ Ⓗ Ⓙ Ⓚ   20. Ⓕ Ⓖ Ⓗ Ⓙ Ⓚ   34. Ⓕ Ⓖ Ⓗ Ⓙ Ⓚ

7. Ⓐ Ⓑ Ⓒ Ⓓ Ⓔ   21. Ⓐ Ⓑ Ⓒ Ⓓ Ⓔ   35. Ⓐ Ⓑ Ⓒ Ⓓ Ⓔ

8. Ⓕ Ⓖ Ⓗ Ⓙ Ⓚ   22. Ⓕ Ⓖ Ⓗ Ⓙ Ⓚ   36. Ⓕ Ⓖ Ⓗ Ⓙ Ⓚ

9. Ⓐ Ⓑ Ⓒ Ⓓ Ⓔ   23. Ⓐ Ⓑ Ⓒ Ⓓ Ⓔ   37. Ⓐ Ⓑ Ⓒ Ⓓ Ⓔ

10. Ⓕ Ⓖ Ⓗ Ⓙ Ⓚ   24. Ⓕ Ⓖ Ⓗ Ⓙ Ⓚ   38. Ⓕ Ⓖ Ⓗ Ⓙ Ⓚ

11. Ⓐ Ⓑ Ⓒ Ⓓ Ⓔ   25. Ⓐ Ⓑ Ⓒ Ⓓ Ⓔ   39. Ⓐ Ⓑ Ⓒ Ⓓ Ⓔ

12. Ⓕ Ⓖ Ⓗ Ⓙ Ⓚ   26. Ⓕ Ⓖ Ⓗ Ⓙ Ⓚ   40. Ⓕ Ⓖ Ⓗ Ⓙ Ⓚ

13. Ⓐ Ⓑ Ⓒ Ⓓ Ⓔ   27. Ⓐ Ⓑ Ⓒ Ⓓ Ⓔ   41. Ⓐ Ⓑ Ⓒ Ⓓ Ⓔ

14. Ⓕ Ⓖ Ⓗ Ⓙ Ⓚ   28. Ⓕ Ⓖ Ⓗ Ⓙ Ⓚ   42. Ⓕ Ⓖ Ⓗ Ⓙ Ⓚ

© 2006 Kaplan, Inc.

43. (A) (B) (C) (D) (E)   59. (A) (B) (C) (D) (E)

44. (F) (G) (H) (J) (K)   60. (F) (G) (H) (J) (K)

45. (A) (B) (C) (D) (E)

46. (F) (G) (H) (J) (K)

47. (A) (B) (C) (D) (E)

48. (F) (G) (H) (J) (K)

49. (A) (B) (C) (D) (E)

50. (F) (G) (H) (J) (K)

51. (A) (B) (C) (D) (E)

52. (F) (G) (H) (J) (K)

53. (A) (B) (C) (D) (E)

54. (F) (G) (H) (J) (K)

55. (A) (B) (C) (D) (E)

56. (F) (G) (H) (J) (K)

57. (A) (B) (C) (D) (E)

58. (F) (G) (H) (J) (K)

© 2006 Kaplan, Inc.

MATHEMATICS TEST
60 Minutes—60 Questions

DIRECTIONS: Solve each problem, choose the correct answer, and then fill in the corresponding oval on your answer document.

Do not linger over problems that take too much time. Solve as many as you can; then return to the others in the time you have left for this test.

You are permitted to use a calculator on this test. You may use your calculator for any problems you choose, but some of the problems may best be done without using a calculator.

Note: Unless otherwise stated, all of the following should be assumed.

1. Illustrations are NOT necessarily drawn to scale.
2. Geometric figures lie in a plane.
3. The word *line* indicates a straight line.
4. The word *average* indicates arithmetic mean.

---

1. A league is a unit of length, equal to 3 miles. If a marathon is 26.2 miles, how many leagues is a marathon, to the nearest tenth?

   A. 78.6
   B. 45.9
   C. 23.2
   D. 13.1
   E. 8.7

2. Because of increased rents in the area, a pizzeria needs to raise the cost of its $20.00 extra large pizza by 22%. What will the new cost be?

   F. $20.22
   G. $22.20
   H. $24.00
   J. $24.40
   K. $42.00

DO YOUR FIGURING HERE.

GO ON TO THE NEXT PAGE.

© 2006 Kaplan, Inc.

3. A group of friends were comparing the number of cards in their baseball card collection. The number of cards is indicated in the table below.

DO YOUR FIGURING HERE.

Friend	A	B	C	D	E
Number of Baseball Cards	30	110	70	200	70

What is the average number of baseball cards in the collection of the 5 friends?

A. 70

B 82

C. 96

D. 102.5

E. 120

4. Train A travels 50 miles per hour for 3 hours; Train B travels 70 miles per hour for $2\frac{1}{2}$ hours. What is the *difference* between the number of miles traveled by Train A and the number of miles traveled by Train B?

F.   0

G.   25

H.   150

J.   175

K.   325

5. Which of the following is a value of b for which $(b-3)(b+4) = 0$?

A.   3

B.   4

C.   7

D.   10

E.   12

GO ON TO THE NEXT PAGE.

© 2006 Kaplan, Inc.

6. In the parallelogram *RSTU*, $\overline{ST}$ is 8 feet long. If the parallelogram's perimeter is 42 feet, how many feet long is $\overline{UT}$?

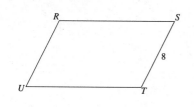

   F. 34

   G. 26

   H. 21

   J. $15\frac{1}{4}$

   K. 13

7. If the measure of each interior angle of a regular polygon is 60°, how many sides does the polygon have?

   A. 3

   B. 4

   C. 6

   D. 10

   E. 12

8. For all nonzero *a*, *b* and *c* values, $\dfrac{12a^5bc^7}{{}^-3ab^5c^2} = ?$

   F. $\dfrac{{}^-4c^5}{a^4b^4}$

   G. $\dfrac{{}^-4a^4c^5}{b^4}$

   H. $\dfrac{{}^-4ac}{b}$

   J. $^-4a^6b^6c^9$

   K. $^-4a^4b^4c^5$

GO ON TO THE NEXT PAGE.

© 2006 Kaplan, Inc.

9. In the figure below, P and Q lie on the sides of $\triangle WXY$, and $\overline{PQ}$ is parallel to $\overline{WY}$. What is the measure of $\angle QPX$?

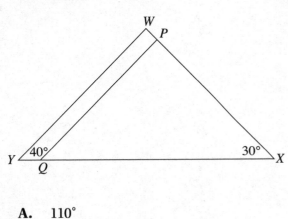

- **A.** 110°
- **B.** 120°
- **C.** 130°
- **D.** 140°
- **E.** 150°

10. $|{^-4}| \cdot |2| = ?$

- **F.** −8
- **G.** −6
- **H.** −2
- **J.** 6
- **K.** 8

11. A class took a vote of their favorite type of music. Of the 32 students, 20 said they liked R&B, 8 said they liked Rock, and 4 students liked Top 40. What percent of the participants liked Rock?

- **A.** 8%
- **B.** 12%
- **C.** 25%
- **D.** 63%
- **E.** 80%

GO ON TO THE NEXT PAGE.

© 2006 Kaplan, Inc.

12. Two whole numbers have a greatest common factor of 15 and a least common multiple of 225. Which of the following pairs of numbers will satisfy this condition?

    F.   9 and 25
    G.   15 and 27
    H.   25 and 45
    J.   30 and 45
    K.   45 and 75

DO YOUR FIGURING HERE.

13. If $x = 2$ and $y = {}^-3$, then $x^5y + xy^5 = $?

    A. $^-60$
    B. $^-192$
    C. $^-390$
    D. $^-582$
    E. $^-972$

14. How many units long is one side of a square with perimeter $20 + 8j$ units?

    F. $20 + 8j$
    G. $20 + 2j$
    H. $4j$
    J. $5 + 8j$
    K. $5 + 2j$

GO ON TO THE NEXT PAGE.

© 2006 Kaplan, Inc.

15. If $(x - k)^2 = x^2 - 26x + k^2$ for all real numbers x, then $k = ?$

DO YOUR FIGURING HERE.

   A. 13

   B. 26

   C. 52

   D. 104

   E. 208

16. Helena bought her daughter a game system and two game cartridges for her birthday, all on sale. The game system, regularly $180, was 10% off, and the game cartridges, regularly $40 each, were 20% off. What was the total price of the 3 items Helena bought?

(Note: Assume there is no sales tax.)

   F. $186

   G. $194

   H. $221

   J. $226

   K. $250

17. Which of the following expressions gives the slope of the line connecting the points $(5, 9)$ and $(^-3, ^-12)$?

   A. $\dfrac{9 + (^-12)}{^-5 - (^-3)}$

   B. $\dfrac{9 + (^-12)}{^-3 + 5}$

   C. $\dfrac{9 - (^-12)}{5 - (^-3)}$

   D. $\dfrac{9 - (^-12)}{^-3 - 5}$

   E. $\dfrac{9 - (^-12)}{^-5 + 3}$

GO ON TO THE NEXT PAGE.

© 2006 Kaplan, Inc.

18. In the standard (x, y) coordinate plane, how many times does the graph of $y = (x - 1)(x + 7)(x - 11)(x + 13)$ intersect the x-axis?

   F.   32
   G.   11
   H.   7
   J.   4
   K.   1

DO YOUR FIGURING HERE.

19. Which of the following is an equivalent, simplified version of $\dfrac{4 + 8x}{12x}$?

   A. $\dfrac{2x + 1}{3x}$
   B. $\dfrac{1 + 8x}{3x}$
   C.   1
   D. $\dfrac{7}{3}$
   E. $\dfrac{8}{3}$

20. Four friends about to share an airport shuttle for $21.50 per ticket discover that they can purchase a book of 5 tickets for $95.00. How much would each of the 4 save if they can get a fifth person to join them and they divide the cost of the book of 5 tickets equally among all 5 people?

   F. $2.25
   G. $2.50
   H. $3.13
   J. $9.00
   K. $12.50

GO ON TO THE NEXT PAGE.

© 2006 Kaplan, Inc.

21. What is the sum of the polynomials
$^-2x^2y^2 + x^2y$ and $3x^2y^2 + 2xy^2$?

   A. $^-6x^4y^4 + 2x^3y^3$

   B. $^-2x^2y^2 + x^2y + 2xy^2$

   C. $x^2y^2 + x^2y + 2xy^2$

   D. $x^2y^2 + x^2y$

   E. $x^2y^2 + 3x^2y$

DO YOUR FIGURING HERE.

22. A 12-foot flagpole casts a 7-foot shadow
when the angle of elevation of the sun is θ
(see figure below). What is $\tan(\theta)$?

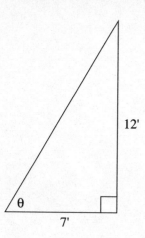

   F. $\dfrac{7}{12}$

   G. 1

   H. $\dfrac{12}{7}$

   J. 19

   K. 84

23. Yousuf was x years old 15 years ago. How old
will he be 7 years from now?

   A. $x + 7$

   B. $(x - 15) + 7$

   C. $(x + 15) - 7$

   D. $(x - 15) - 7$

   E. $(x + 15) + 7$

GO ON TO THE NEXT PAGE.

© 2006 Kaplan, Inc.

24. Which of the following is a factor of
    $x^2 - 4x - 12$?

    F. $(x + 1)$
    G. $(x - 2)$
    H. $(x + 2)$
    J. $(x - 3)$
    K. $(x - 4)$

DO YOUR FIGURING HERE.

25. What is the length, in inches, of the hypotenuse of a right triangle with legs measuring 8 inches and 15 inches?

    A. 7
    B. 17
    C. 23
    D. $\sqrt{23}$
    E. $\sqrt{161}$

26. Which of the following expressions is a simplified form of $(^-2x^5)^3$?

    F. $^-6x^8$
    G. $8x^8$
    H. $^-2x^{15}$
    J. $^-6x^{15}$
    K. $^-8x^{15}$

GO ON TO THE NEXT PAGE.

© 2006 Kaplan, Inc.

27. If 1 mile is approximately equal to 1.609 kilometers, how many miles are in 80 kilometers?

    A. 49.7

    B. 78.4

    C. 80

    D. 81.6

    E. 128.7

28. If $2x + 3 = {}^-5$, what is the value of $x^2 - 7x$?

    F. $^-44$

    G. $^-12$

    H. $^-4$

    J. 12

    K. 44

29. Which of the following is a graph of the solution set for $2(5 + x) < 2$?

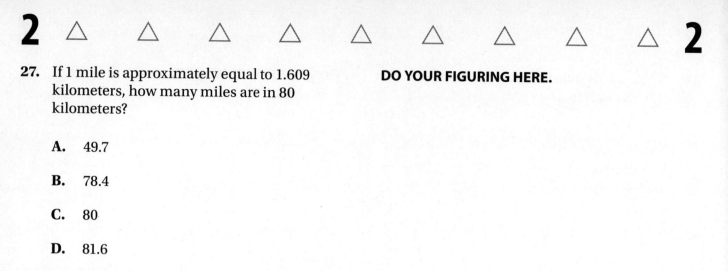

    A.

    B.

    C.

    D.

    E.

GO ON TO THE NEXT PAGE.

© 2006 Kaplan, Inc.

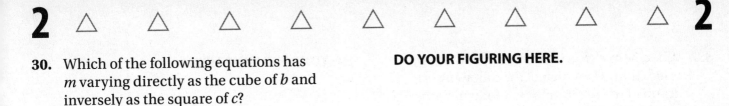

DO YOUR FIGURING HERE.

30. Which of the following equations has *m* varying directly as the cube of *b* and inversely as the square of *c*?

    F. $\dfrac{m^3}{c^2} = b$

    G. $\dfrac{b^3}{c^2} = m$

    H. $\dfrac{c^3}{b^2} = m$

    J. $\dfrac{\sqrt[3]{b}}{c} = m$

    K. $\dfrac{b^3}{m^2} = c$

31. Points $V(^-2,^-7)$ and $W(4, 5)$ determine line segment $\overline{VW}$ in the standard (x, y) coordinate plane. If the midpoint of $\overline{VW}$ is $(1, p)$, what is the value of *p*?

    A. $^-2$

    B. $^-1$

    C. 1

    D. 2

    E. 6

32. If the graphs of $y = 3x$ and $y = mx + 6$ are parallel in the standard (x,y) coordinate plane, then what is the value of *m*?

    F. $^-6$

    G. $\dfrac{1}{3}$

    H. 2

    J. 3

    K. 6

GO ON TO THE NEXT PAGE.

© 2006 Kaplan, Inc.

33. When 3 times x is increased by 5, the result is less than 11. Which of the following is a graph of the real numbers x for which the previous statement is true?

DO YOUR FIGURING HERE.

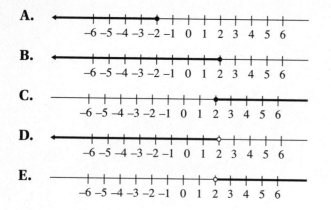

A.

   −6 −5 −4 −3 −2 −1 0 1 2 3 4 5 6

B.

   −6 −5 −4 −3 −2 −1 0 1 2 3 4 5 6

C.

   −6 −5 −4 −3 −2 −1 0 1 2 3 4 5 6

D.

   −6 −5 −4 −3 −2 −1 0 1 2 3 4 5 6

E.

   −6 −5 −4 −3 −2 −1 0 1 2 3 4 5 6

34. It costs 54 cents to buy x pencils and 92 cents to buy y erasers. Which of the following is an expression for the cost, in cents, of 7 pencils and 3 erasers?

F. $\dfrac{54}{7+x} + \dfrac{92}{3+y}$

G. $3\left(\dfrac{54}{x}\right) + 7\left(\dfrac{92}{y}\right)$

H. $7\left(\dfrac{x}{54}\right) + 3\left(\dfrac{y}{92}\right)$

J. $7\left(\dfrac{54}{x}\right) + 3\left(\dfrac{92}{y}\right)$

K. $7\left(\dfrac{92}{x}\right) + 3\left(\dfrac{54}{x}\right)$

35. When graphed in the standard (x,y) coordinate plane, 3 points from among $(^-9,^-7)$, $(^-5,^-3)$, $(^-2,^-1)$, $(1,^-1)$ and $(10,^-8)$ lie on the same side of the line $y - x = 0$. Which of the three points are they?

A. $(^-9, ^-7)$, $(^-2, ^-1)$, $(^-5, ^-3)$

B. $(^-9, ^-7)$, $(^-2, ^-1)$, $(1, ^-1)$

C. $(^-9, ^-7)$, $(^-5, ^-3)$, $(10, ^-8)$

D. $(^-9, ^-7)$, $(1, ^-1)$, $(10, ^-8)$

E. $(^-5, ^-3)$, $(1, ^-1)$, $(10, ^-8)$

GO ON TO THE NEXT PAGE.

© 2006 Kaplan, Inc.

36. What is the sine of angle E in right triangle $\triangle DEF$ below?

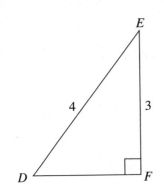

F. $\dfrac{\sqrt{7}}{3}$

G. $\dfrac{3}{4}$

H. $\dfrac{\sqrt{7}}{4}$

J. $\dfrac{3}{\sqrt{7}}$

K. $\dfrac{4}{\sqrt{7}}$

37. The graph of the solution set for the system of linear equations below is a single line in the (x,y) coordinate plane.

$$12x - 20y = 108$$

$$3x + ky = 27$$

What is the value of k?

A. $^{-}5$

B. $^{-}3$

C. $\dfrac{1}{4}$

D. $\dfrac{3}{5}$

E. 4

GO ON TO THE NEXT PAGE.

© 2006 Kaplan, Inc.

38. A common rule of thumb is that each additional inch of height (H) will add 10 pounds to a person's weight (W). Doctors recommend finding your Body Mass Index (BMI) as a measure of health. BMI is computed as follows (H is in inches, and W is in pounds):

$$BMI = \frac{703W}{H^2}$$

If a 68-inch-tall person typically weighs 150 pounds, which of the following is closest to the expected BMI of a 72-inch-tall person?

F.   1

G.   2

H.   20

J.   26

K.   42

DO YOUR FIGURING HERE.

39. Dave's math tutor reminded him not to calculate $\left(\dfrac{x}{y}\right)^2$ as $\dfrac{x^2}{y}$. Dave thinks there are some numbers for which that calculation works. Eventually, he was able to show that $\left(\dfrac{x}{y}\right)^2$ equals $\dfrac{x^2}{y}$ if and only if:

(Note: Assume that $y \neq 0$.)

A. $x = 0$

B. $x = 1$

C. $y = 1$

D. $x = 0$ and $y = 1$

E. $x = 0$ or $y = 1$

GO ON TO THE NEXT PAGE.

© 2006 Kaplan, Inc.

40. In the figure below, $\overline{BD}$ is a perpendicular bisector of $\overline{AC}$ in equilateral triangle $\triangle ABC$. If $\overline{BD}$ is $4\sqrt{3}$ units long, how many units long is $\overline{BC}$?

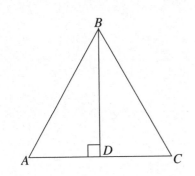

F. $2\sqrt{3}$

G. 4

H. 8

J. $8\sqrt{3}$

K. 16

41. What is the perimeter, in meters (m), of the figure below?

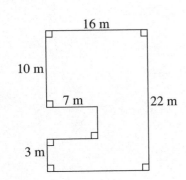

A. 58

B. 83

C. 90

D. 208

E. 352

GO ON TO THE NEXT PAGE.

© 2006 Kaplan, Inc.

42. Isosceles trapezoid *ABCD* is inscribed in a circle with center *O*, as shown below. Which of the following is the most direct explanation of why △*AOD* is isosceles?

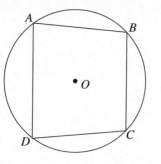

F. If two angles in a triangle are congruent, the sides opposite them are congruent.

G. 2 sides are radii of the circle.

H. Side-angle-side congruence

J. Angle-side-angle congruence

K. Angle-angle-angle similarity

43. A circle with radius 4 meters is cut out of a circle with radius 12 meters, as shown in the figure below. Which of the following gives the area of the shaded figure, in square meters?

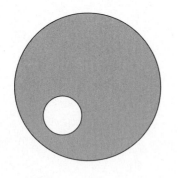

A. $\pi(12-2)^2$

B. $\pi 12^2 - 2^2$

C. $\pi 12^2 - 4^2$

D. $\pi(12-4^2)$

E. $\pi(12^2 - 4^2)$

GO ON TO THE NEXT PAGE.

© 2006 Kaplan, Inc.

DO YOUR FIGURING HERE.

44. A walkway, 31 by $32\frac{1}{2}$ feet, surrounds a pool that is $27\frac{1}{2}$ by 29 feet, as shown below.

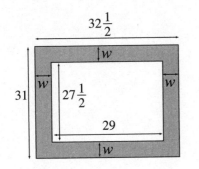

What is the width, w, in feet, of the walkway?

F. $1\frac{1}{4}$

G. 1

H. $1\frac{1}{2}$

J. $1\frac{3}{4}$

K. $3\frac{1}{2}$

GO ON TO THE NEXT PAGE.

© 2006 Kaplan, Inc.

45. The area of a rectangular floor is 323 square feet. The width of the floor is 21 feet less than twice the length. How many feet long is the floor?

    A.   8.5

    B.   11

    C.   13.5

    D.   17

    E.   19

DO YOUR FIGURING HERE.

46. For the area of a circle to double, the new radius must be the old radius multiplied by:

    F. $\dfrac{1}{2}$

    G. $\sqrt{2}$

    H.   2

    J. π

    K.   4

GO ON TO THE NEXT PAGE.

© 2006 Kaplan, Inc.

47. If $\log_x 64 = 3$, then $x = ?$

    A. 4

    B. 8

    C. $\dfrac{64}{3}$

    D. $\dfrac{64}{\log 3}$

    E. 64^3

48. If $A = \begin{bmatrix} 3 & ^{-}6 \\ 0 & 9 \end{bmatrix}$ and $B = \begin{bmatrix} ^{-}3 & 6 \\ 0 & ^{-}9 \end{bmatrix}$ then $A - B = ?$

    F. $\begin{bmatrix} 0 & 0 \\ 0 & 0 \end{bmatrix}$

    G. $\begin{bmatrix} 1 & 0 \\ 0 & 1 \end{bmatrix}$

    H. $\begin{bmatrix} 0 & ^{-}12 \\ 0 & 18 \end{bmatrix}$

    J. $\begin{bmatrix} ^{-}6 & 0 \\ 0 & 0 \end{bmatrix}$

    K. $\begin{bmatrix} 6 & ^{-}12 \\ 0 & 18 \end{bmatrix}$

49. If a and b are real numbers, and $a > 0$ and $b < a$, then which of the following inequalities must be true?

    A. $b \leq 0$

    B. $b \geq 0$

    C. $b^2 \geq 0$

    D. $b^2 \geq a^2$

    E. $b^2 \leq a^2$

GO ON TO THE NEXT PAGE.

© 2006 Kaplan, Inc.

50. The ratio of the lengths of the sides of a right triangle is $2 : \sqrt{5} : 3$. What is the cosine of the smallest angle in the triangle?

DO YOUR FIGURING HERE.

F. $\dfrac{2}{3}$

G. $\dfrac{\sqrt{5}}{3}$

H. $\dfrac{2\sqrt{5}}{5}$

J. $\dfrac{9}{10}$

K. 2

51. What is the amplitude of the graph of the equation $y + 3 = 4\sin(5\theta)$?

(Note: the amplitude is $\dfrac{1}{2}$ the difference between the maximum and the minimum values of y.)

A. 3

B. 4

C. 5

D. 7

E. 10

52. Each of the following determines a unique plane in 3-dimensional Euclidian space EXCEPT:

F. 1 line and 1 point NOT on the line.

G. 3 distinct points NOT on the same line.

H. 2 lines that intersect in exactly 1 point.

J. 2 distinct parallel lines.

K. 2 lines that are NOT parallel and do NOT intersect.

GO ON TO THE NEXT PAGE.

© 2006 Kaplan, Inc.

2 △ △ △ △ △ △ △ △ △ 2

53. The measure of the vertex angle of an isosceles triangle is $(x - 10)°$. The base angles each measure $(3x + 18)°$. What is the measure in degrees of one of the base angles?

A. 12

B. 22

C. $37\frac{1}{2}$

D. $43\frac{1}{2}$

E. 84

54. To make a set of potholders of various sizes to give as a gift, Margot needs the following amounts of fabric for each set:

pieces of fabric	length (inches)
6	8
5	12
2	18

If the fabric costs $1.95 per yard, which of the following would be the approximate cost of fabric for 5 sets of potholders?

(Note: 1 yard = 36 inches)

F. $ 8

G. $ 24

H. $ 39

J. $ 58

K. $ 117

DO YOUR FIGURING HERE.

GO ON TO THE NEXT PAGE.

© 2006 Kaplan, Inc.

55. The formula for the surface area (S) of a rectangular solid with square bases (shown below) is $S = 4wh + 2w^2$, where w is the side length of the bases, and h is the height of the solid. Doubling each of the dimensions (w and h) will increase the surface area to how many times its original size?

DO YOUR FIGURING HERE.

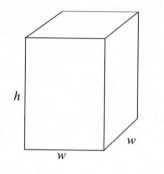

- **A.** 2
- **B.** 4
- **C.** 6
- **D.** 8
- **E.** 24

56. The average of a set of four integers is 14. When a fifth number is included in the set, the average of the set increases to 16. What is the fifth number?

- **F.** 16
- **G.** 18
- **H.** 21
- **J.** 24
- **K.** 26

GO ON TO THE NEXT PAGE.

ACT ADVANTAGE
MATHEMATICS

© 2006 Kaplan, Inc.

57. Which of the following is the equation of the largest circle that can be inscribed in the ellipse with equation $\dfrac{(x-4)^2}{16} + \dfrac{y^2}{4} = 1$?

   A. $(x-4)^2 + y^2 = 64$
   B. $(x-4)^2 + y^2 = 16$
   C. $(x-4)^2 + y^2 = 4$
   D. $x^2 + y^2 = 16$
   E. $x^2 + y^2 = 4$

DO YOUR FIGURING HERE.

GO ON TO THE NEXT PAGE.

© 2006 Kaplan, Inc.

58. One of the graphs below is that of $y = x^3 + C$, where C is a constant. Which one?

DO YOUR FIGURING HERE.

F.

G.

H.

J.

K.

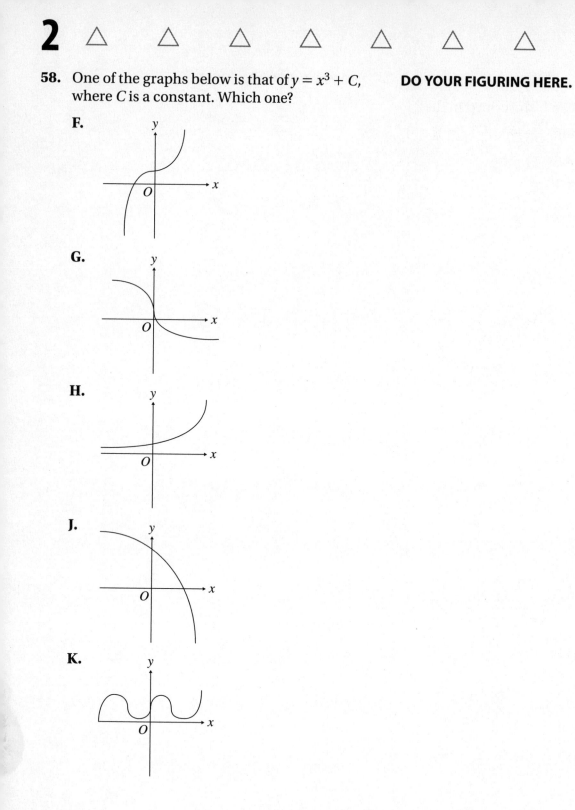

GO ON TO THE NEXT PAGE.

© 2006 Kaplan, Inc.

59. How many points do the graphs of all three equations below have in common?

$x = y + 8$

$^-x = y - 8$

$6x = 2y + 4$

A. 0

B. 1

C. 2

D. 3

E. An infinite amount

DO YOUR FIGURING HERE.

GO ON TO THE NEXT PAGE.

© 2006 Kaplan, Inc.

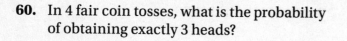

60. In 4 fair coin tosses, what is the probability of obtaining exactly 3 heads?

(Note: In a fair coin toss, the two outcomes, heads and tails, are equally likely.)

F. $\dfrac{1}{16}$

G. $\dfrac{1}{8}$

H. $\dfrac{3}{16}$

J. $\dfrac{1}{4}$

K. $\dfrac{1}{2}$

DO YOUR FIGURING HERE.

END OF TEST.

STOP! DO NOT TURN THE PAGE UNTIL TOLD TO DO SO.

© 2006 Kaplan, Inc.

Math Reference

© 2006 Kaplan, Inc.

Rules and Formulas

Geometry

Type of Polygon	Relevant Formulas
Rectangle (including square)	Area = (*Length*)(*Width*)
Parallelogram (including rhombus)	*Area = (Base)(Height)*
Circle	Area = πr^2 Circumference = $2\pi r$

Type of Solid	Relevant Formulas
Cube	Total Surface Area = $6s^2$ Volume = s^3
Right Circular Cylinder	Total Surface Area = $2\pi rh + 2\pi r^2$ Volume = $\pi r^2 h$
Right Rectangular Prism	Total Surface Area = $2w\ell + 2\ell h + 2wh$ Volume = ℓwh
Right Triangular Prism	Total Surface Area = $w\ell + wh + \ell h + \ell s$ Volume = $\frac{1}{2}w \times h \times \ell$

© 2006 Kaplan, Inc.

Coordinate Geometry

Term	Definition	How To Find It
Slope	the ratio of the change in y over the change in x (this describes the rise over run, or the slant of the line)	1. Pick two points on the line. 2. Find the difference between the y- and x-coordinates and place them in a ratio. $$m = \dfrac{y_2 - y_1}{x_2 - x_1}$$
x-intercept	the point at which the line crosses the x-axis	1. Set $y = 0$ in the equation. 2. Solve for x. The resulting value is the x-intercept.
y-intercept	the point at which the line crosses the y-axis	1. Set $x = 0$ in the equation. 2. Solve for y. The value found is the y-intercept.

Trigonometry

$$\sin\theta = \frac{opposite}{hypotenuse}$$

$$\cos\theta = \frac{adjacent}{hypotenuse}$$

$$\tan\theta = \frac{\sin\theta}{\cos\theta} = \frac{opposite}{adjacent}$$

$$\sec\theta = \frac{1}{\cos\theta} = \frac{hypotenuse}{adjacent}$$

$$\cot\theta = \frac{1}{\tan\theta} = \frac{adjacent}{opposite}$$

$$\csc\theta = \frac{1}{\sin\theta} = \frac{hypotenuse}{opposite}$$

© 2006 Kaplan, Inc.

Exponents

Rules	Examples
$x^0 = 1$	$v^0 = 1$ $3^0 = 1$
$x^m \cdot x^n = x^{m+n}$	$z^2 \cdot z^3 = z^5$ $2^2 \times 2^3 = 2^5 = 32$
$x^1 = x$	$5^1 = 5$
$\dfrac{x^m}{x^n} = x^{m-n}$	$\dfrac{b^8}{b^2} = b^6$ $\dfrac{5^4}{5} = 5^3 = 125$
$x^{-n} = \dfrac{1}{x^n}$	$g^{-4} = \dfrac{1}{g^4}$ $7^{-2} = \dfrac{1}{7^2} = \dfrac{1}{49}$
$(x^m)^n = x^{mn}$	$(p^3)^3 = p^9$ $(3^2)^3 = 3^6 = 729$
$(xy)^n = x^n y^n$	$(jk)^8 = j^8 k^8$ $(4r)^2 = 4^2 \cdot r^2 = 16r^2$
$\left(\dfrac{x}{y}\right)^n = \dfrac{x^n}{y^n}$	$\left(\dfrac{a}{b}\right)^6 = \dfrac{a^6}{b^6}$ $\left(\dfrac{2}{3}\right)^2 = \dfrac{2^2}{3^2} = \dfrac{4}{9}$

© 2006 Kaplan, Inc.

Sequences

Sequence	Rule
arithmetic sequence	Each new number is obtained from the previous number by adding or subtracting a constant.
geometric sequence	Each new number is obtained from the previous number by multiplying by a constant.
repeating pattern	A pattern with a string of terms that repeats.
visual patterns	Each new shape is obtained by observing how the figures in the pattern are changing.

© 2006 Kaplan, Inc.

Glossary

Absolute value: A number's distance from zero on a number line.

Acute angle: An angle whose measure is between 0° and 90°.

Acute triangle: A triangle with three acute angles.

Addend: A number being added to obtain a sum.

Additive inverse: Also called the opposite of a number. The sum of any number and its additive inverse is zero. $a + (-a) = 0$.

Algebra: A generalization of the concepts of arithmetic. In algebra, unknown numbers are called *variables* and are usually represented by letters.

Algebraic equation: A mathematical sentence that includes one or more variables and contains two expressions joined by an equal sign.

Algebraic expression: A combination of numbers and variables with one or more operations. For example, $x, x + 7, 3x - 5$, and $2x^2 - 3y^2$.

Algebraic rule: An expression containing variables that represents a relationship or pattern.

Alternate interior angles: When parallel lines are cut by a transversal, angles between the parallel lines and on opposite sides on the transversal are called alternate interior angles, and are congruent.

Angle: A figure formed by two rays with a common endpoint. The common endpoint is called the vertex of the angle.

Angle bisector: A ray that divides an angle into two smaller, congruent angles.

Area: The number of square units needed to cover a region.

Axes: The vertical and horizontal lines of reference that divide the coordinate plane into quadrants. The horizontal axis is the x-axis and the vertical axis is the y-axis.

Axis (axes): A reference line used to read information on a graph.

Bar graph: A type of graph using vertical or horizontal bars to indicate relationships among data.

Base: The number in an exponential expression that is used as a factor. For example, in 3^2, 3 is the base.

Bases of a prism: Two congruent and parallel faces of a prism.

Break: A zigzag or squiggle on the vertical or horizontal axis of a graph showing that the graph omits a range of values on the scale being used.

Capacity: The amount of available space in a three-dimensional region, usually associated with measuring liquid or dry material. Volume and capacity are often used interchangeably.

Central angle: An angle that has its vertex at the midpoint of a circle.

Chart: A display used to organize information or data.

Circle graph: A type of graph where a circle represents the total amount; it is also known as a pie chart.

Circumference: The distance around a circle.

© 2006 Kaplan, Inc.

Closed figure: A plane figure in which the beginning and endpoints touch. (Circles, regular polygons, and irregular polygons are all closed figures.)

Common factor: A common factor of two natural numbers is a number that is a factor of both numbers. For example, 7 is a common factor of 35 and 56, because $35 = 7 \times 5$ and $56 = 7 \times 8$.

Common multiple: A common multiple of two natural numbers is a number that is a multiple of both numbers. For example, 42 is a common multiple of 6 and 14 because $42 = 6 \times 7$ and $42 = 14 \times 3$.

Complementary angles: A pair of angles, the sum of whose measures is 90°

Composite number: An integer greater than 1 with more than two factors.

Cone: A solid with a circular base and a vertex that is not in the same plane as the base.

Congruent figures: Two or more figures that are the same shape and size.

Consecutive numbers: Whole numbers that follow in order, such as 3, 4, 5.

Consecutive sides of a polygon: Adjacent sides of a polygon.

Constant: A symbol representing a value that does not change. For example, 5, π, and $\sqrt{2}$ are constants.

Convert: To express a given measurement in terms of a different unit of measurement.

Coordinate grid plot: A plot that uses an ordered pair to denote location on the grid. Points are located first by the horizontal coordinate and then by the vertical coordinate.

Coordinate system: A plane formed by a horizontal line (the x-axis) that intersects a vertical line (the y-axis). The plane is divided into four quadrants in which ordered pairs, or points, are identified and plotted.

Coordinates: The ordered pair (x, y) used to represent a point on a coordinate plane.

Corresponding angles in parallel lines cut by a transversal: When parallel lines are cut by a transversal, angles in the same position relative to the parallel lines and transversal are corresponding and congruent.

Corresponding angles of similar triangles: In similar triangles, the angles in the same relative location of the two triangles are corresponding and congruent. If $\triangle ABC \sim \triangle XYZ$, then $\angle A \cong \angle X$ and $m\angle A = m\angle X$.

Corresponding sides of similar triangles: In similar triangles, the sides in the same relative location of the two triangles are corresponding and proportional. If $\triangle ABC \sim \triangle XYZ$, then sides AB, BC, AC are in proportion to the corresponding sides XY, YZ, XZ.

Cosine: In a right triangle, the ratio of the length of the side adjacent to a given angle to the length of the hypotenuse.

Cube: A rectangular prism with congruent edges and faces.

Cubed: Another way to say "to the third power."

Cubic units: The units used to measure the space inside a three-dimensional figure.

Customary system of measurement: The system of measurement used in the United States. The basic units include the foot, gallon, and pound.

Cylinder: A solid with congruent and parallel circular bases.

Data: Information about a situation, group, or event.

Data set: A collection of number facts.

© 2006 Kaplan, Inc.

Decimal: A number written in standard notation, containing a decimal point.

Denominator: The number of equal parts into which the whole or group is divided. In the fraction $\frac{a}{b}$, b is the denominator.

Density: The ratio of the mass of an object to its volume.

Diagonal: A line segment that connects non-adjacent vertices in a polygon.

Diameter: A line segment connecting two points on a circle and passing through the center of the circle. The diameter is equal to twice the length of the radius.

Difference: The result of subtracting two numbers.

Digit: Any of the numerals 0 through 9.

Distance: The measurement between two points in a plane.

Distributive Property: The product of a factor and a sum is equal to the sum of the products. $a(b + c) = ab + ac$

Dividend: The amount to be divided. In the example $a \div b = c$, a is the dividend.

Divisible: An integer is divisible by another integer when the division leaves a remainder of zero.

Division: The inverse operation of multiplication. The operation is performed on two numbers to obtain a third number, called the quotient.

Division laws of exponents: To divide two powers that have the same base, subtract the exponents.

For example, $\frac{4^5}{4^3} = 4^2$.

Divisor: The amount by which a number is divided. In the example $a \div b = c$, b is the divisor.

Double-bar graph: A graph with bars representing two different sets of data.

Double-line graph: A graph with two lines representing different sets of data.

Edge: The line or line segment along which two faces of a solid figure meet.

Element: An item in a set.

Endpoints: The initial point and terminal point of a line segment.

Equation: A mathematical sentence that uses an equal sign to show that two quantities are equal.

Equilateral triangle: A triangle with sides of equal length and angles of equal measure. All of the angles in an equilateral triangle have a measure of 60°.

Equivalent fractions: Fractions that name the same number.

Equivalent ratios: Ratios that make the same comparison.

Event: A subset of the possible outcomes of an experiment.

Experiment: A controlled study in which subjects are divided into groups and an action is taken on one or more of the groups. The reactions to the treatment are recorded.

Experimental probability: Probability of an event based on the statistical results of an actual experiment that has already been performed. May also be called empirical probability.

Exponent: A symbol written above and to the right of a base that tells how many times the base is used as a factor. For example, $3 \cdot 3 \cdot 3 \cdot 3$ is written 3^4; the exponent is 4.

Exponential form: A way to write numbers using exponents. For example, 2^5 is written in exponential form.

Expression: A mathematical arrangement of numbers and variables connected by

© 2006 Kaplan, Inc.

addition, subtraction, multiplication, or division.

Extrapolate: To interpret a given set of data in order to estimate values outside the known range.

Face: A flat side of a solid figure.

Factor (noun): A number that divides exactly into another number with no remainder. When two or more factors are multiplied, they form a product. For example, $2 \cdot 5 = 10$, 2 and 5 are factors, and 10 is the product.

Factor (verb): To write a number or expression as a product of its factors.

Factored form: A way to write numbers using factors. For example, 3×3 is the factored form of 9.

Factor tree: A graphic tool used to determine the prime factors of a number.

False statement: A mathematical statement that represents a false relationship. For example, $3 > 4$ is a false statement.

Fraction: A number in the form $\frac{a}{b}$, where a and b are integers and b is not zero. Fractions are used to name parts of a whole object or part of a whole collection of objects.

Function: A mathematical relationship between two sets of numbers in which every element of the first set is matched with exactly one element of the second set.

Function table: Pairs of x- and y-values arranged in a table that represent a function, relationship, sequence, or pattern.

Fundamental Counting Principle: A general rule that states that if one event has p possible outcomes and another event has q possible outcomes, then the first event followed by the second event has $p \cdot q$ possible outcomes.

Geometry: The study of the properties, measurement, and relationships of points, lines, angles, planes, and solids.

Gram: The basic unit of mass in the metric system.

Graph: A data display using ordered pairs, bars, lines, circles, or pictures.

Greater than or equal to: The symbol $\geq$. In the inequality $x \geq 5$, x could be greater than 5 or x could be equal to 5.

Greatest common factor: The GCF of two natural numbers is the largest number that is a factor of both numbers. For example, the GCF of 180 and 84 is 12 since $180 = 12 \times 15$ and $84 = 12 \times 7$.

Height of a prism or cylinder: The perpendicular distance between the bases.

Height of a pyramid or cone: The length of the perpendicular segment from the vertex to the base.

Horizontal axis: The axis running left to right, often denoted as the x-axis, in a two-dimensional coordinate system.

Hypotenuse: The side opposite the right angle in a right triangle. The hypotenuse is the longest side of a right triangle.

Hypothesis: A statement formulated to serve as the grounds of subsequent inquiry.

Impossible event: An event that cannot happen. It has a probability of 0.

Independent events: Two or more events in which the outcome of any one event does not affect the outcome of any of the other events.

Inequality: A mathematical sentence that uses symbols such as $>$ (greater than) and $<$ (less than) to compare values.

© 2006 Kaplan, Inc.

Input: Numbers to which a function is applied.

Integer: A whole number, or the opposite of a whole number. Zero is also an integer and is neither positive nor negative. $\{...-3, -2, -1, 0, 1, 2, 3, ...\}$

Interest : A sum of money paid when borrowing or saving money. Interest is a percent of the principal amount and is found using the formula:
Interest = Principal × rate × time.

Interior angles of a triangle: The three angles on the inside of a triangle. The sum of their measures is 180°.

Intersection: The point formed by two lines that cross.

Interval: The evenly-spaced lines or points along the axes of a graph.

Inverse operations: Operations that are opposites. For example, addition is the inverse operation of subtraction and multiplication is the inverse operation of division.

Irrational number: A number that cannot be written in the form of $\frac{a}{b}$ where a and b are integers and b is non-zero. Irrational numbers have neither terminating decimals nor repeating decimals.

Isosceles triangle: A triangle with two congruent sides and two congruent angles.

Lateral edge: A line formed by the intersection of two lateral faces of a prism.

Lateral face: The faces of a prism that are not bases.

Least Common Multiple: The LCM of two natural numbers is the smallest number that is a multiple of both. For example, the LCM of 24 and 30 is 120.

Legend: A key to go with a graph.

Legs: In a right triangle, the two sides that intersect to form the right angle.

Length: A dimension of a rectangle. It usually refers to the longer side.

Less than or equal to: The symbol $\leq$. In the inequality $x \leq 7$, x could be less than 7 or x could be equal to 7.

Like terms: Expressions that have the same variables with the same powers of those variables.

Line: A set of points that form a straight path extending to infinity in both directions.

Line graph: A graph with a horizontal and vertical axis that represents data as a continuous line or curve.

Line segment: A section of a line with a definite beginning and end.

Linear equation: An equation that represents a straight line.

Liter: The basic unit for measuring capacity within the metric system, where one liter equals 1000 milliliters.

Lower extreme: The lowest value in a set of data.

Lower quartile: The lower fourth of a set of data.

Maximum: The largest value in a data set.

Mean: The average of a set of numbers.

Measures of central tendency: Values that summarize a set of numerical data. The most common measures of central tendency are mean, median, and mode.

Median: The middle number in a set of numbers that are arranged in order from least to greatest. When there is an even number of numbers, the median is the average of the two middle numbers.

Metric system: A system of measurement that is based on units of 10. The basic

© 2006 Kaplan, Inc.

units of the metric system are the meter, liter, and gram.

Midpoint: The point on a line segment that is halfway between the endpoints. The midpoint divides the line segment into two equal segments.

Minimum: The smallest value in a data set.

Mixed number: A number consisting of a whole number and a fraction, such as $4\frac{1}{2}$.

Mode: The number that occurs most often in a set of numbers.

Monomial: A polynomial with one term.

Multiple: A multiple of a number is a product of that number and any non-zero whole number.

Multiplication laws of exponents: To multiply two powers that have the same base, add the exponents. For example, $a^2 \times a^3 = a^5$. To find the power of a power, multiply exponents. For example, $(5^2)^4 = 5^8$. To find the power of a product, find the power of each factor and multiply. For example, $(2 \times 3)^6 = 2^6 \times 3^6$.

Multiplication: An operation on two numbers, called factors, to obtain a third number, called the product.

Multiplicative inverse: The opposite of a number in a multiplication operation. For a non-zero number b, the multiplicative inverse is $\frac{1}{b}$, such that $b \times \frac{1}{b} = 1$. The multiplicative inverse is also called the reciprocal.

Natural numbers: The set of positive integers, not including zero. Natural numbers are also known as counting numbers. {1, 2, 3, 4, 5, ...}

Negative exponent: When used in scientific notation, a negative exponent indicates a number less than 1. For example, 6.77×10^{-3} equals 0.00677.

Negative integer: An integer that is less than zero. {$^-1$, $^-2$, $^-3$, ...}

Number line: A line on which points are marked off at regular intervals (i.e., evenly spaced) and labeled with ordered numbers.

Numerator: The number of equal parts being considered when a whole is divided into equal parts. In the fraction $\frac{a}{b}$, a is the numerator.

Numerical expression: A mathematical phrase that contains numbers and operation symbols, but no variables.

Obtuse angle: An angle whose measure is between 90° and 180°.

Obtuse triangle: A triangle with one obtuse angle.

Operation: A process performed on numbers and expressions. The basic operations are addition, subtraction, multiplication, and division, but raising to a power and finding a square root are also included.

Opposite integers: Two integers that are the same distance from zero on a number line in different directions.

Order of operations: A rule indicating the order in which operations should be performed when there are several operations in an expression. Perform operations inside grouping symbols first, then simplify exponents; next, perform multiplication and division from left to right. Finally, perform addition and subtraction from left to right.

Ordered pair: A pair of numbers (x, y) used to name a point on a coordinate plane.

Origin: The point (0, 0) located in a two-dimensional coordinate system. It is the intersection of the x-axis and the y-axis.

Outcome: A result of an experiment.

© 2006 Kaplan, Inc.

Output: Numbers resulting from a function being applied.

Parallel lines: Lines that lie in the same plane and never intersect.

Parallelogram: A four-sided polygon with opposite sides parallel and congruent.

Pattern: A configuration, design, or arrangement that is specific and predictable.

Percent: A special ratio in which the comparison is made out of 100.

Percent decrease: The percent that represents $\frac{amount\ of\ decrease}{original\ amount}$.

Percent increase: The percent that represents $\frac{amount\ of\ increase}{original\ amount}$.

Perfect square: A whole number that can be renamed as the square of an integer.

Perimeter: The distance around a polygon. This distance can be found by adding the lengths of the sides.

Perpendicular lines: Two lines that intersect to form right angles.

Pi (π): The symbol used to represent the ratio of the circumference of a circle to its diameter. It can be approximated as 3.1416.

Pie chart: A type of graph that is also known as a circle graph. The sectors of the circle show proportions of data.

Pint: A dry or liquid measure of capacity. Two cups equal one pint; two pints equal one quart.

Place value: The number assigned to each place occupied by a digit.

Plane: A two-dimensional geometric object that extends without boundary.

Plane figure: A figure that lies in only one plane. Circles and polygons are plane figures.

Plot: To graph ordered pairs (x, y) on the coordinate plane.

Point: A geometric object with position, but neither length nor width.

Polygon: A closed figure with three or more straight sides.

Polynomial: A mathematical expression that contains one or more monomials. For example, $16y^2 + 8y + 4$ is a polynomial.

Positive integer: An integer that is greater than zero. {1, 2, 3, ...}

Power: Another name for exponent.

Prediction: A guess about the outcome or result of an event or action based on the available information.

Prime factorization: The expression of a composite number as the product of its prime factors. For example, the prime factorization of 315 is $3^2 \times 5 \times 7$.

Prime number: An integer greater than 1 with only two factors: 1 and itself.

Principal: A sum of money increased by a rate of interest.
Interest = Principal $\times$ rate $\times$ time.

Prism: A solid formed by three or more lateral faces that intersect at bases.

Probability: The likelihood or chance of an event occurring, expressed as a number from 0 to 1. A probability of 1 means the event is certain, while a probability of 0 means the event cannot occur. Probability is equal to the ratio of the number of favorable outcomes to the number of total outcomes.

Product: The result of multiplication.

Property of Proportions: In the proportion $\frac{a}{b} = \frac{c}{d}$, $c \times b = a \times d$.

© 2006 Kaplan, Inc.

Proportion: An equation stating that two ratios are equivalent.

Protractor: A tool used for measuring angles in degrees.

Pyramid: A solid formed by the intersection of a polygonal base with triangular faces that meet at a vertex.

Pythagorean theorem: The sum of the squares of the lengths of the legs in a right triangle is equal to the square of the length of the hypotenuse. In a right triangle with a and b as the lengths of the legs and c as the length of the hypotenuse, this relationship can be expressed as $a^2 + b^2 = c^2$.

Quadrant: One of the four sections, or quarters, located on a coordinate plane.

Quart: A unit of liquid measure. Two pints equal one quart, and four quarts equal one gallon.

Quartile: A number marking $\frac{1}{4}$ of a set of ordered numbers.

Quotient: The result of division. In the example $a \div b = c$, c is the quotient.

Radical sign: The symbol $\sqrt{}$ used to denote a square root.

Radicand: The expression under a radical sign.

Radius: A line segment connecting a point on a circle with the center of the circle. The radius is half the length of the diameter.

Range: The difference between the largest number and the smallest number in a set.

Rate: A comparison of two quantities that have different units of measure. For example, miles per hour, cans per dollar, miles per gallon, and dollars per pound.

Ratio: A comparison of two numbers or quantities. It may be written as a fraction.

Rational number: A number that can be expressed in the form $\frac{a}{b}$, where a and b are integers and b is not equal to zero.

Ray: A portion of a line that extends from a given point in one direction only.

Real numbers: The set of all rational and irrational numbers.

Reciprocal: The reciprocal of a number is equal to 1 divided by the original number. A number multiplied by its reciprocal equals 1.

Regular polygon: A convex polygon with sides of equal length and angles of equal measure.

Regular pyramid: A pyramid where the base has congruent sides and congruent angles and the height meets the base at its center.

Relative size: The size of a number or magnitude in relation to one or more other numbers or magnitudes.

Relevant information: Pertinent or useful information for solving a problem.

Remainder: A whole number that is left over after one whole number is divided by another.

Repeated multiplication in exponential form: A way to write numbers in exponential form. For example, 2^3 can be written as $2 \times 2 \times 2$.

Repeating decimal: A decimal with a block of one or more digits that repeats without end. For example, the fraction $\frac{2}{11}$ can be expressed as the repeating decimal $0.\overline{18}$.

Right angle: An angle whose measure is exactly $90°$.

Right cone: A cone in which the height meets the center of the base.

© 2006 Kaplan, Inc.

Right cylinder: A cylinder in which the height meets the center of both bases.

Right rectangular prism: A prism with rectangular bases that are perpendicular to the lateral faces.

Right triangle: A triangle with one 90° angle.

Rise: The vertical distance, or change in y, between two points on a coordinate grid.

Rounding: Arriving at an approximation by following a prescribed algorithm.

Rule: A set of operations that corresponds to a pattern or relationship. A rule can be represented as a mathematical expression or written description.

Run: The horizontal distance, or change in x, between two points on a coordinate grid.

Sample space: The set of all possible outcomes for an experiment.

Scale: An arrangement of numbers that increase by equal intervals. There may be a scale for the horizontal axis, for the vertical axis, for both axes, or for neither axis.

Scale drawing: A reduced or enlarged drawing whose shape is the same as the actual object and whose size is determined by a scale factor.

Scale factor: The ratio used to compare measurements on a scale drawing with measurements of the actual object.

Scalene triangle: A triangle that does not have any congruent sides.

Scientific notation: A form of writing numbers as a product of a base number that has an absolute value greater than or equal to 1 and less than 10, and a power of 10.

Sequence: The order in which things or events occur or are arranged.

Set: A collection of objects or events.

Side: One of the line segments that enclose a polygon.

Similar figures: Figures that have the same shape but not necessarily the same size. Corresponding angles are congruent, and corresponding side lengths are proportional.

Sine: In a right triangle, the ratio of the length of the side opposite a given angle to the length of the hypotenuse.

Slope: The slant or steepness of a line graphed on the coordinate plane. It is a ratio that compares the change in the y-values (the rise) to the change in the x-values (the run). Lines that rise up from left to right have positive slopes. Lines that fall down from left to right have negative slopes.

Solid figure: A closed three-dimensional figure. Prisms, spheres, cylinders, and pyramids are examples of solid figures.

Solving: The process of determining the answer to a problem.

Solving an equation: Finding the value of the variable that makes an equation true.

Speed: Distance divided by time.

Square root: One of the two equal factors of a number. Geometrically, the square root is the length of the side of a square that has an area equal to the number for which you are taking the square root. For example, $\sqrt{16} = 4$ because a square with an area of 16 square units has sides measuring 4 units. The symbol for square root is $\sqrt{}$.

Square units: The units used to measure the area of a planar region.

Squared: Another way to say "raised to the second power."

© 2006 Kaplan, Inc.

Standard form: The way numbers are usually written. For example, in the expression $3^2 = 9$, 9 is written in standard form and 3^2 is written in exponential form.

Statistics: The study of collecting, organizing, and interpreting data.

Straight angle: An angle whose measure is exactly 180°.

Subtraction: The inverse operation of addition. An operation is performed on two numbers to obtain a third number, called the difference. In the equation $6 - 2 = 4$, the number 4 is the difference.

Substitution: When a variable, expression, or a constant is used in place of another. Substitution is most often used when a variable is replaced by a constant.

Sum: The result of adding numbers.

Sum of the interior angles of a polygon: The total number of degrees in all angles inside of a polygon. The sum of the interior angles of a regular polygon is equal to $180°(n - 2)$, where n is the number of sides of that polygon.

Supplementary angles: A pair of angles the sum of whose measures is 180°.

Surface area: The area of the surface of a figure.

Symbolic representation of numbers: A non-numeric means used to represent numbers. A variable is an example of symbolic representation.

Symmetry: The characteristic of a figure that can be divided into congruent sections such that the sections are mirror images of each other.

Table: A display of information or data in which the facts can be easily read or understood.

Temperature conversion: A correlation between degrees Fahrenheit and degrees Celsius given by the formulas: $F = \frac{9}{4}C + 32$ and $C = \frac{5}{9}(F - 32)$. For example, $0°C = 32°F$.

Tangent: In a right triangle, the ratio of the length of the side opposite a given angle to the length of the side adjacent to the angle.

Terminating decimal: A decimal that ends without repeating. The fraction $\frac{3}{4}$ can be expressed as the terminating decimal 0.75.

Theorem: A formula, proposition, or statement that can be deduced from other formulas.

Theoretical probability: Probability that is expressed as the ratio of the number of times an event can happen to the total number of possible outcomes. This is a probability based on what is likely to occur if an experiment takes place.

Transversal: A line that intersects two or more coplanar lines.

Trapezoid: A four-sided polygon with one pair of parallel sides.

Tree diagram: A diagram showing all the possible combinations of elements or outcomes of an event.

Triangle: A three-sided polygon. The sum of the interior angles is 180°.

True statement: A mathematical statement that represents a relationship that is true. For example, $3 < 4$ is a true statement.

Truth value: The determination of whether a statement is true or false. For example, $3 > 2$ has a truth value of *true*.

Unit price: The price per unit. For example, if 5 bottles cost $1.25, the unit price is $0.25 per bottle.

© 2006 Kaplan, Inc.

Universal set: The set of all objects considered.

Unlikely: The description of a result that may occur, but probably will not.

Upper extreme: The greatest value in a set of data.

Upper quartile: The upper fourth of a set of values.

Variable: An unknown quantity. It is often represented by a letter.

Variable expression: A mathematical phrase that contains at least one variable.

Venn diagram: A visual device used to represent the relationships among sets.

Vertex: The point shared by two rays or line segments forming an angle. It can also refer to the point at which two lines intersect.

Vertical angles: Pairs of opposite congruent angles formed by intersecting lines.

Vertical axis: The axis running top to bottom, often denoted as the y-axis, in a two-dimensional coordinate system.

Volume: The number of cubic units inside, or the capacity of, a three-dimensional figure. Volume formula for a prism or cylinder: $V = Bh$ where B = area of base and h = height of the prism or cylinder.

Whole numbers: The set of natural numbers and zero. {0, 1, 2, 3, 4, …}

Width: A dimension of a rectangle. It usually refers to the shorter side.

x-axis: The horizontal axis in a two-dimensional coordinate system.

x-intercept: The x-coordinate at which a crosses (or intercepts) the x-axis.

y-axis: The vertical axis in a two-dimensional coordinate system.

y-intercept: The y-coordinate at which a line crosses (or intercepts) the y-axis. A line can be identified by its slope and y-intercept in slope-intercept form: $y = mx + b$.

Zero exponent: Any term with zero as the exponent is equal to one.

Zero Property of Multiplication: If a and b are real numbers and $ab = 0$, then $a = 0$ or $b = 0$.

© 2006 Kaplan, Inc.